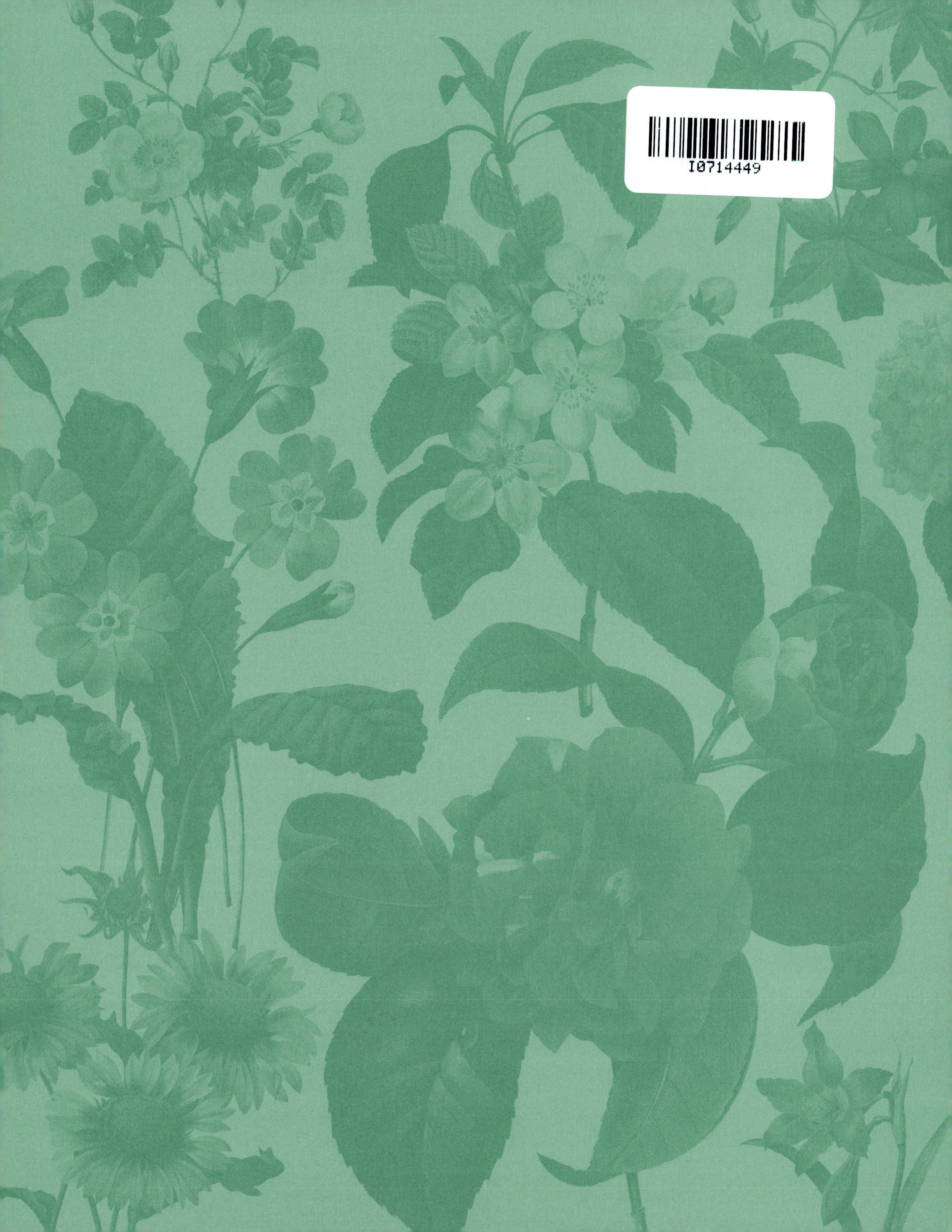

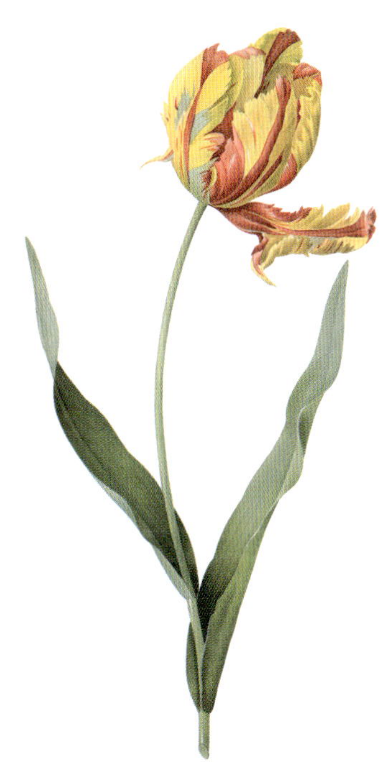

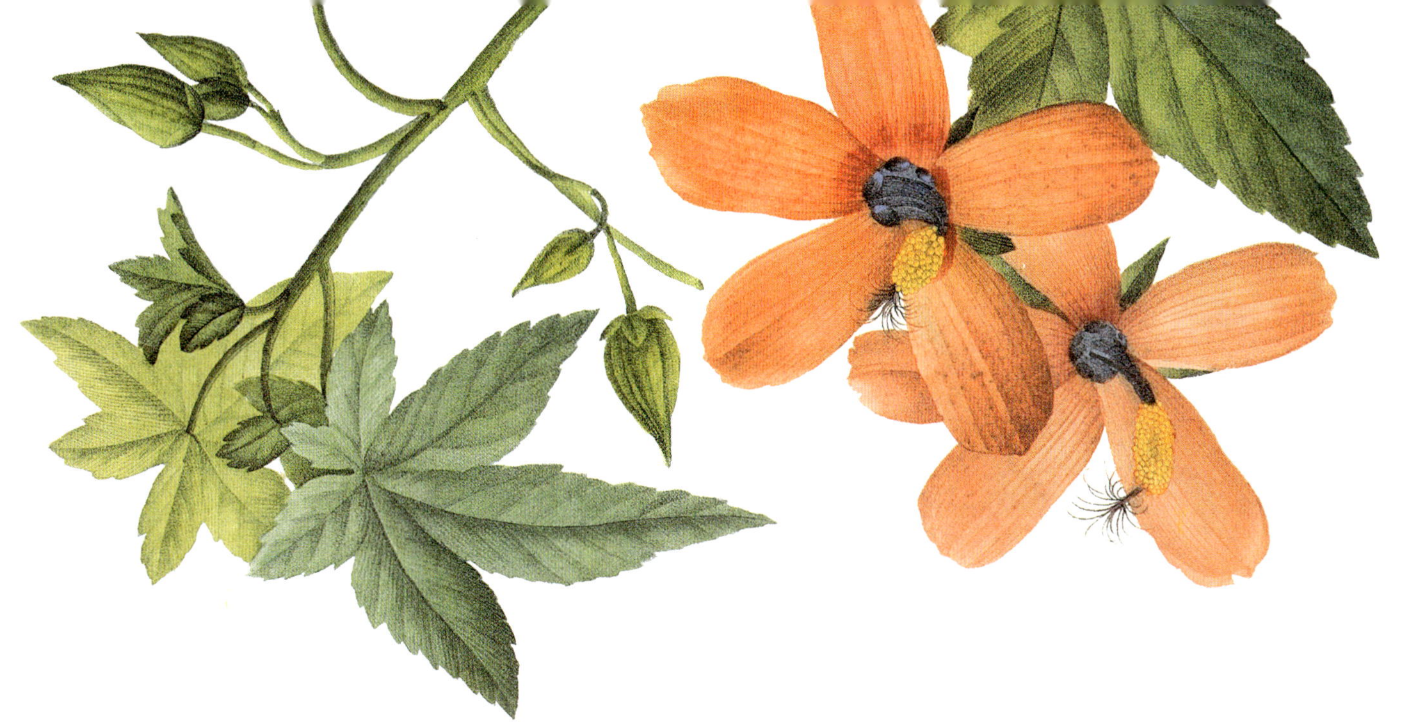

REDOUTÉ
FABULOUS FLOWERS

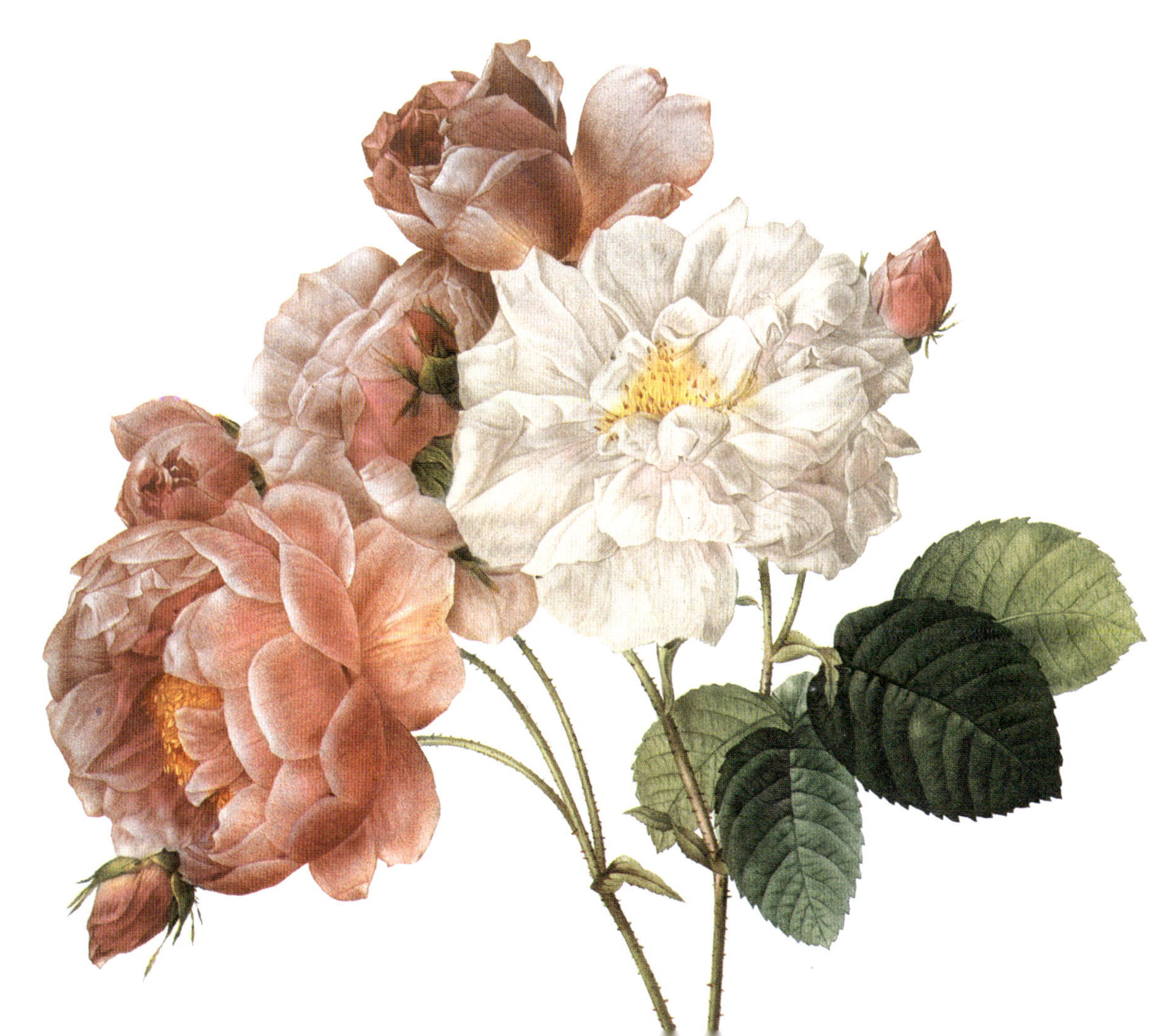

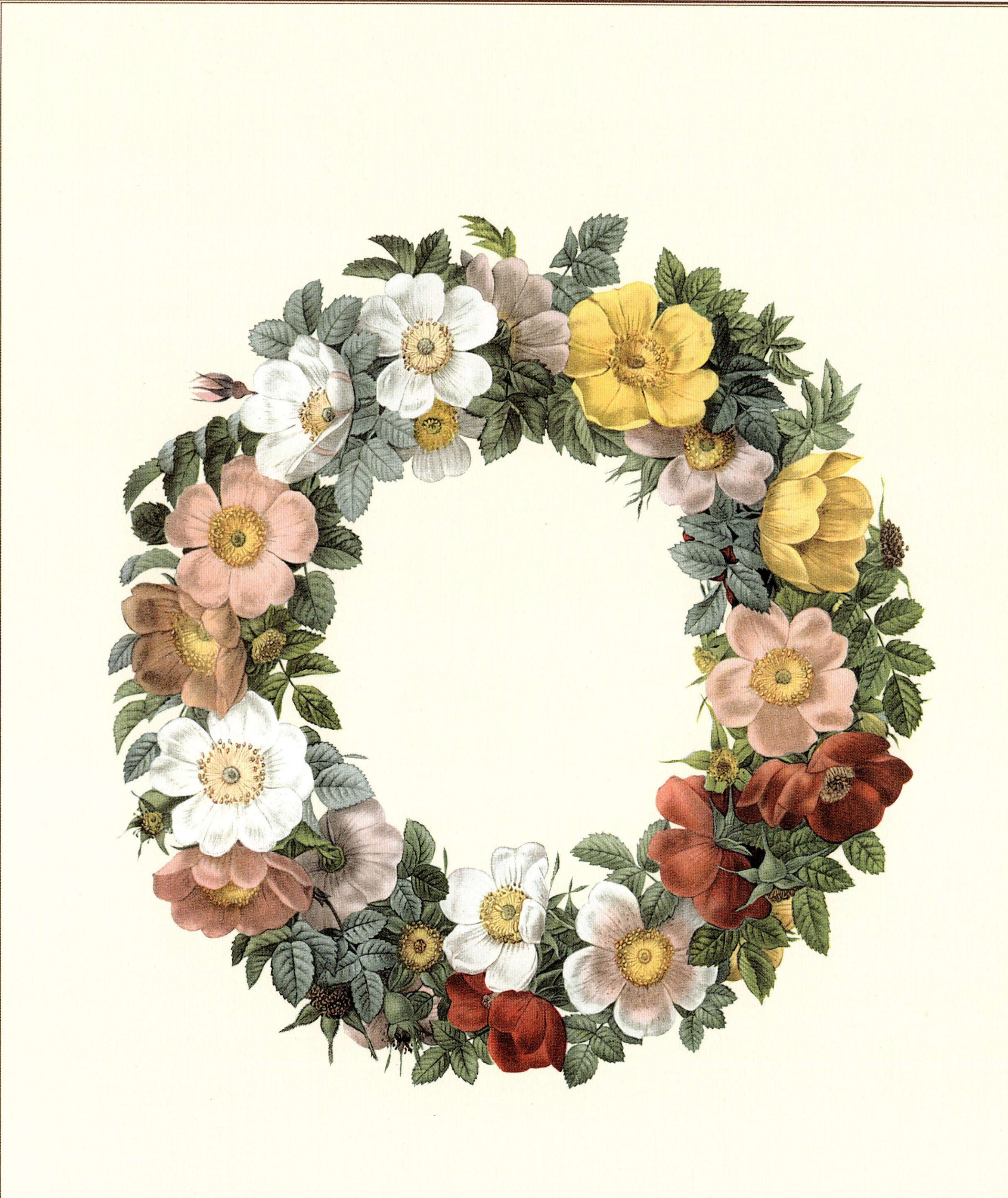

REDOUTÉ
FABULOUS FLOWERS

PIERRE-JOSEPH REDOUTÉ

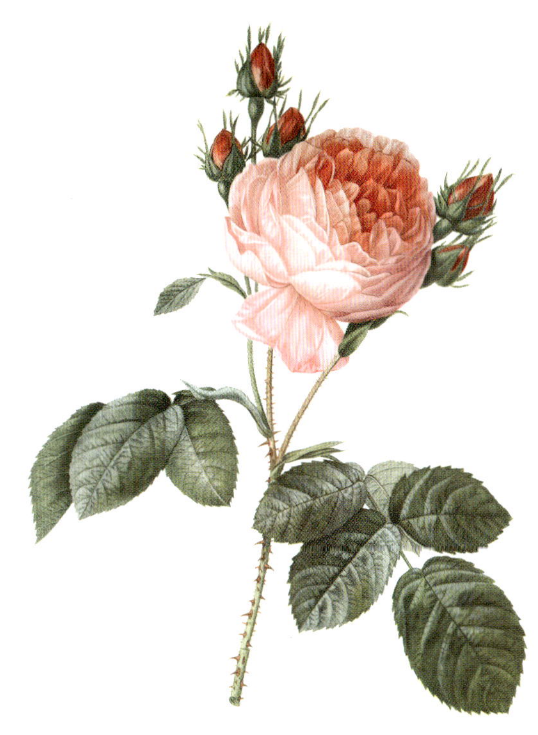

DOVER PUBLICATIONS
GARDEN CITY, NEW YORK

Redouté: Fabulous Flowers is a new work, first published by Dover Publications in 2024.
The plates are reproduced from standard editions of Pierre-Joseph Redouté's collections.

ISBN-13: 978-0-486-85406-9
ISBN-10: 0-486-85406-X

Publisher: Betina Cochran
Associate Publisher: Peter Lenz
Managing Editorial Supervisor: Susan Rattiner
Production Editor: Gregory Koutrouby
Cover Designer: Peter Donahue
Creative Manager and Interior Designer: Marie Zaczkiewicz
Production: Pam Weston, Tammi McKenna, Ayse Yilmaz

Manufactured in China
85406X01   2024
www.doverpublications.com

NOTE

Pierre-Joseph Redouté (1759–1840) is one of the most accomplished botanical painters, known especially for his depictions of roses and lilies. In his lifetime, he was called the Raphael of Flowers. He is to flower paintings what John James Audubon is to bird paintings. Redouté created more than 2,100 paintings featuring more than 1,800 species.

Redouté was born in the Netherlands (in present-day Belgium) and lived most of his life in France. He tutored Marie Antoinette, the last Queen of France, and was named Draughtsman and Painter to the Queen's Cabinet. Empress Josephine, the first wife of Napoleon Bonaparte, was his patron and had him paint flowers in the garden at their estate. Louise of Orléans, the Queen of Belgium, was also his pupil. Redouté received the Legion of Honor medal in France and was knighted in Belgium.

He found favor with powerful admirers, across borders and despite political upheaval. His meticulously accurate watercolors impressed botanists as well as fine art aficionados. Today his botanical prints are beloved in framed pictures, books, and china patterns.

Flowers are the epitome of natural beauty. Redouté captured, celebrated, and championed their beauty, drawing attention to many varieties of flowers that would otherwise go unappreciated. Captions include the English common name and the Latin species name for ninety different types of flowers.

LIST OF PLATES

PLATE

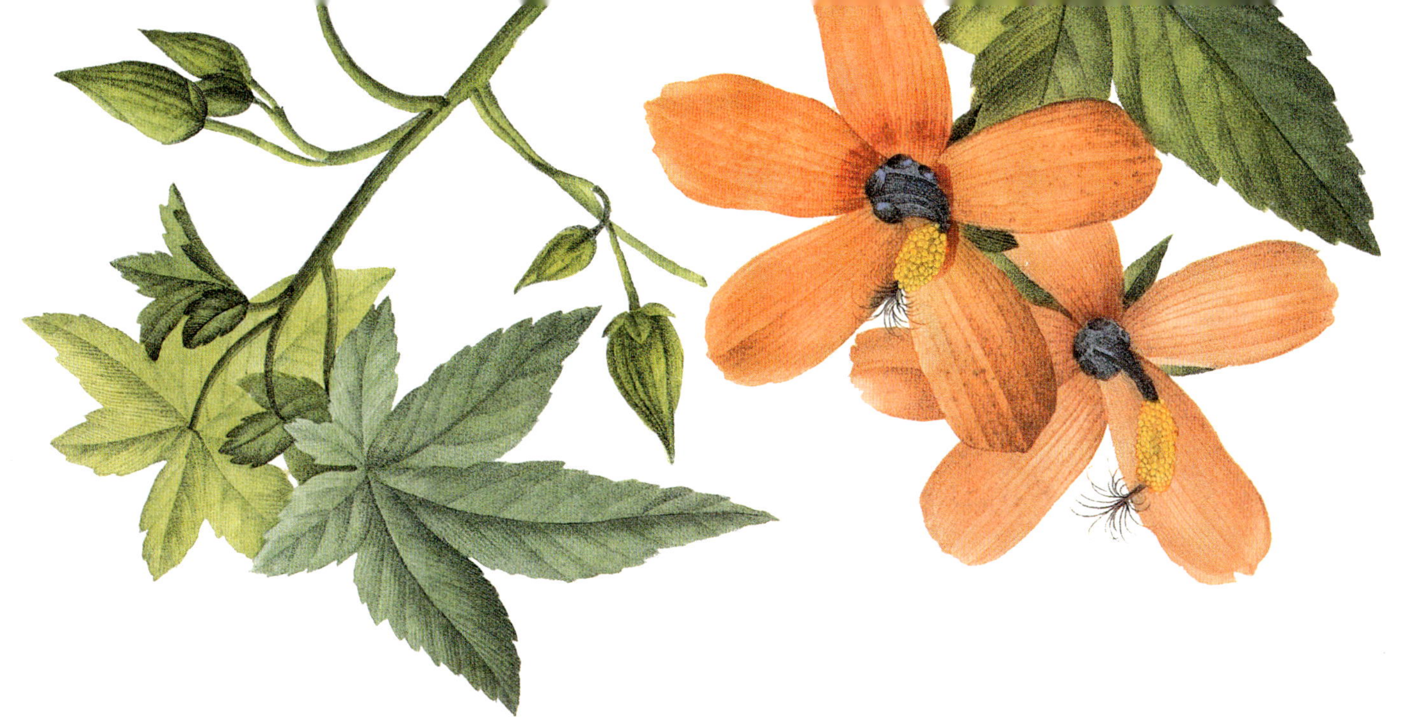

REDOUTÉ
FABULOUS FLOWERS

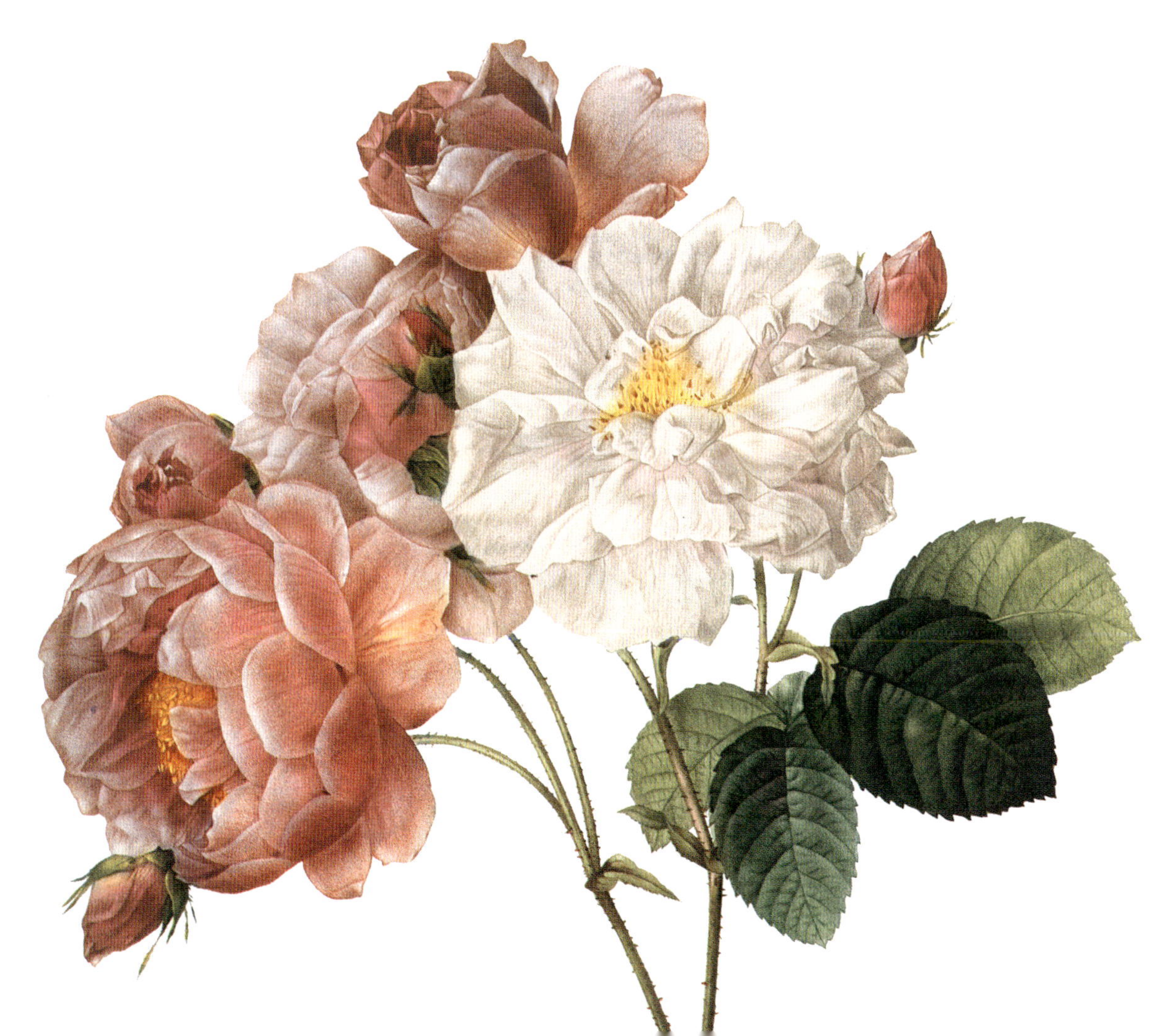

PLATE 1

ALPINE ROSE WITH GLOBOSE RECEPTACLE AND GLABROUS PEDICEL
(Rosa alpina laevis)

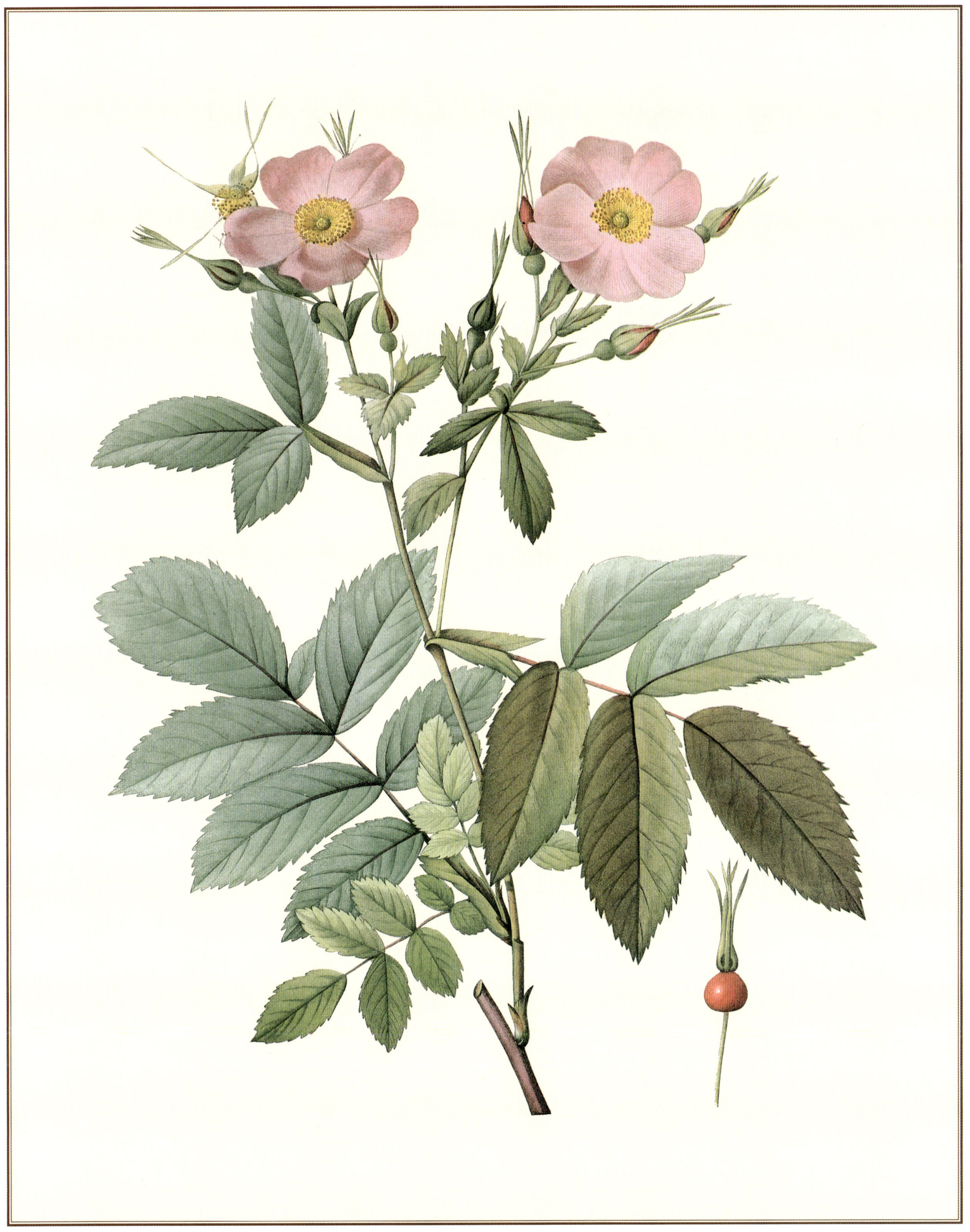

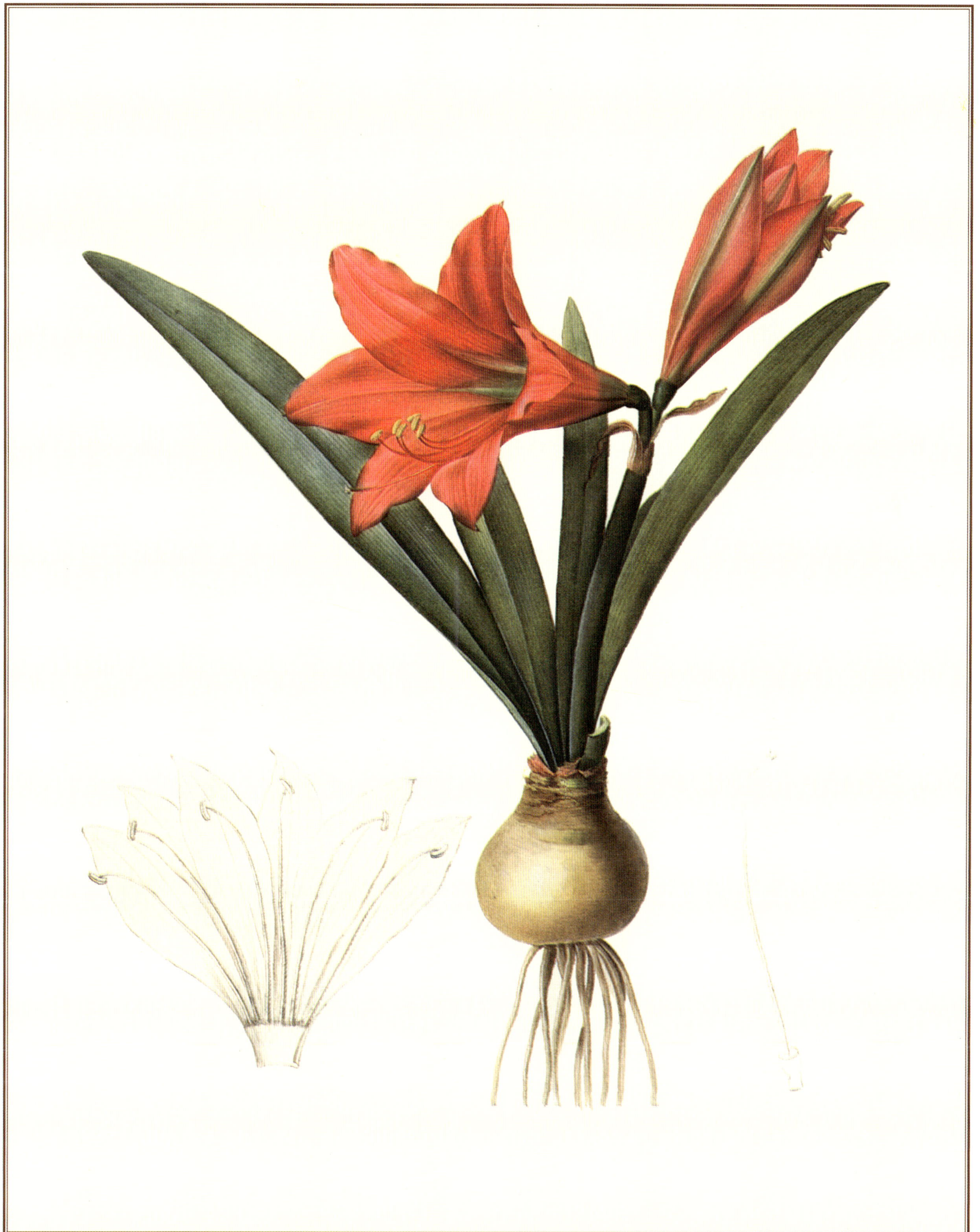

PLATE 3
ANEMONE
(*Anemone coronaria*)

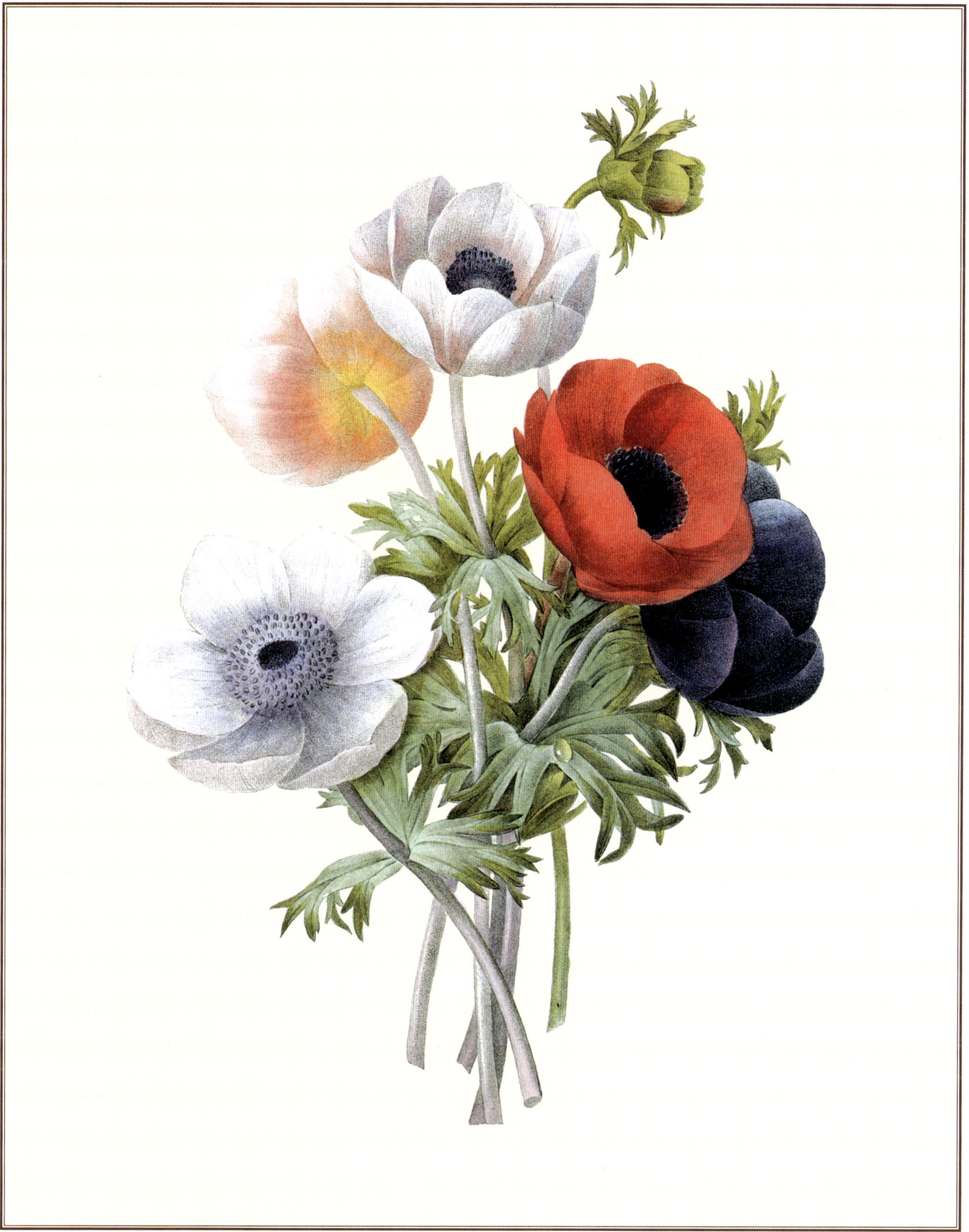

PLATE 4
ANJOU ROSE
(*Rosa andegavensis*)

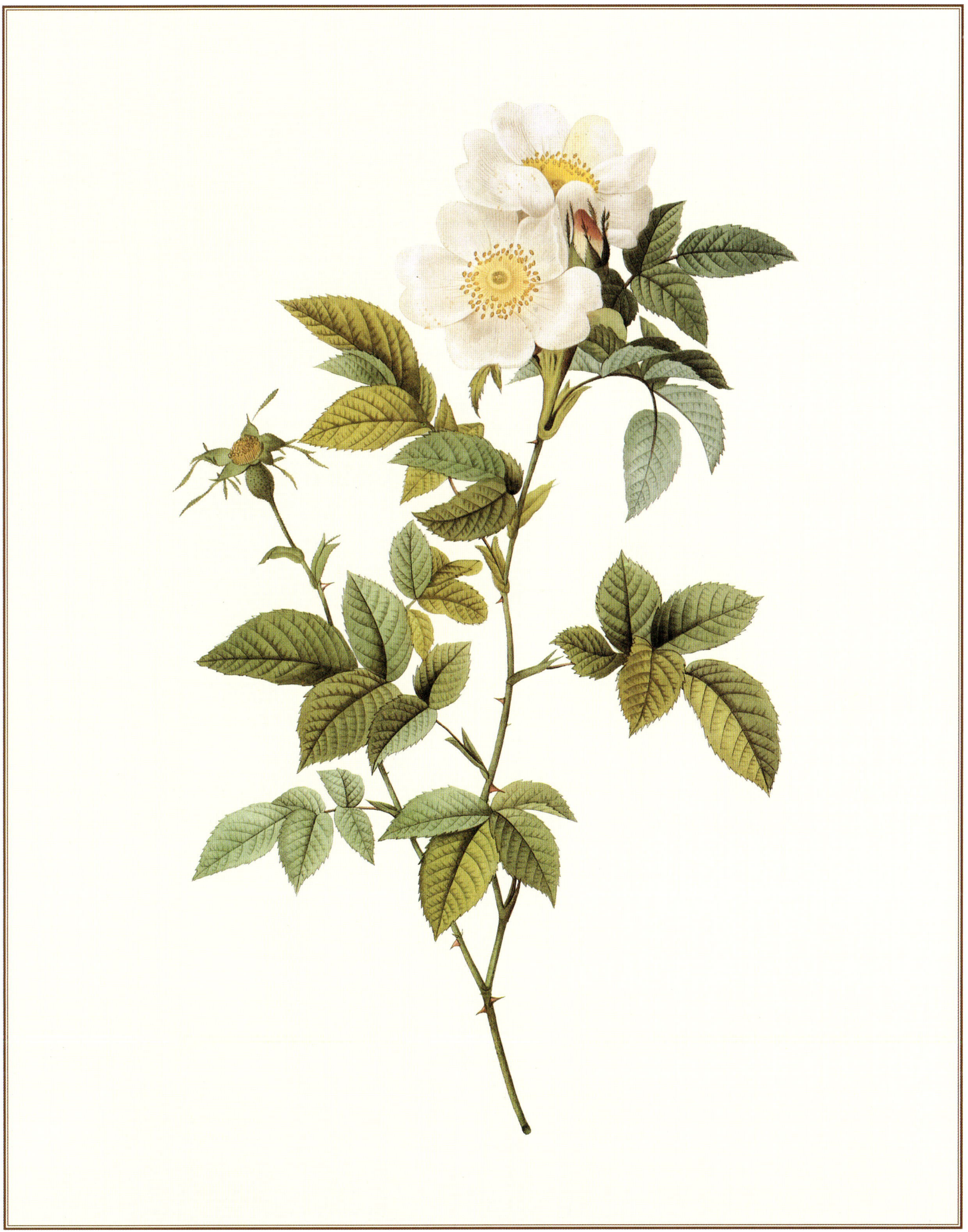

Plate 5

Apple Blossom

(Malus domestica)

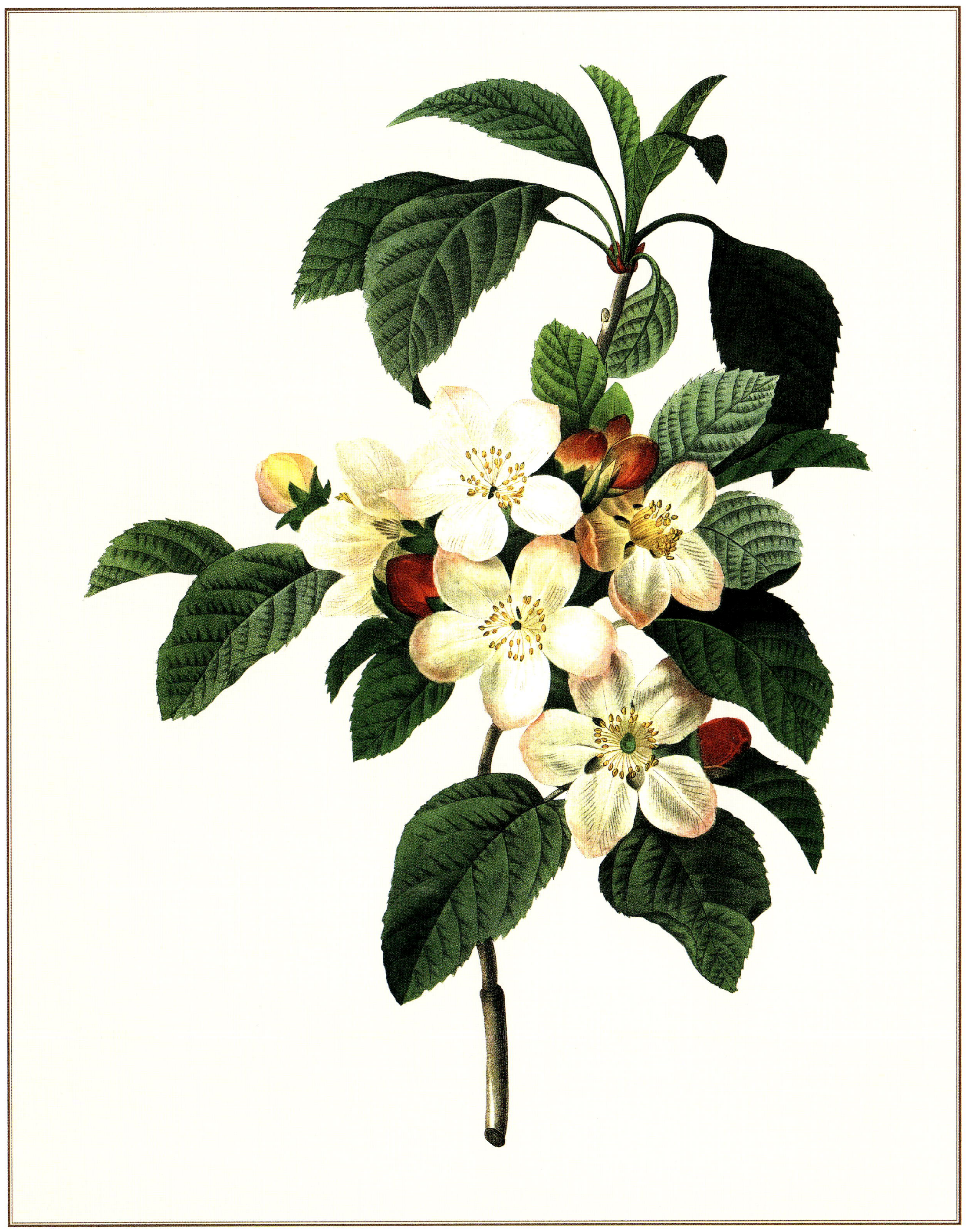

PLATE 6
AUSTRIAN COPPER (CORN POPPY ROSE)
(*Rosa eglanteria punicea*)

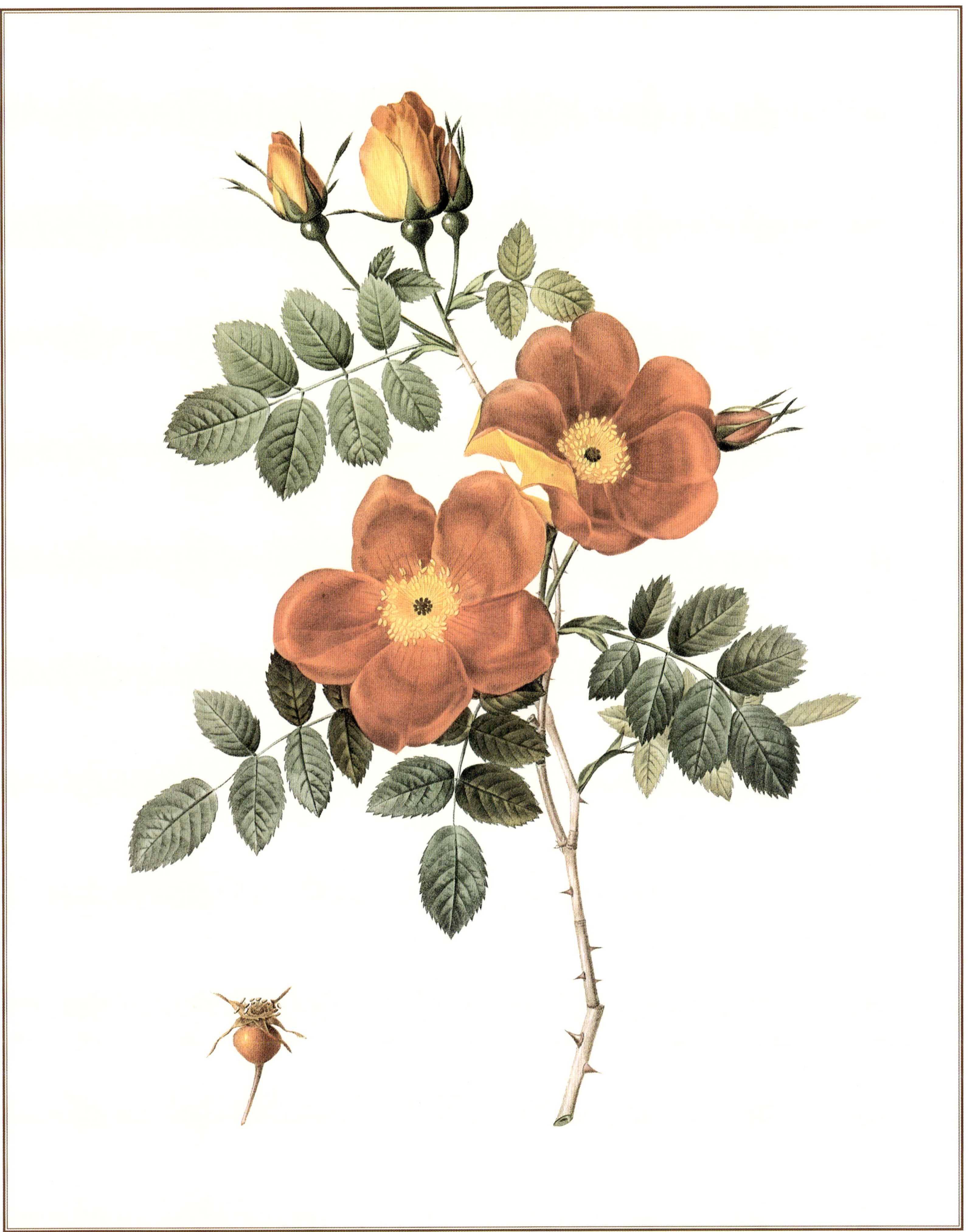

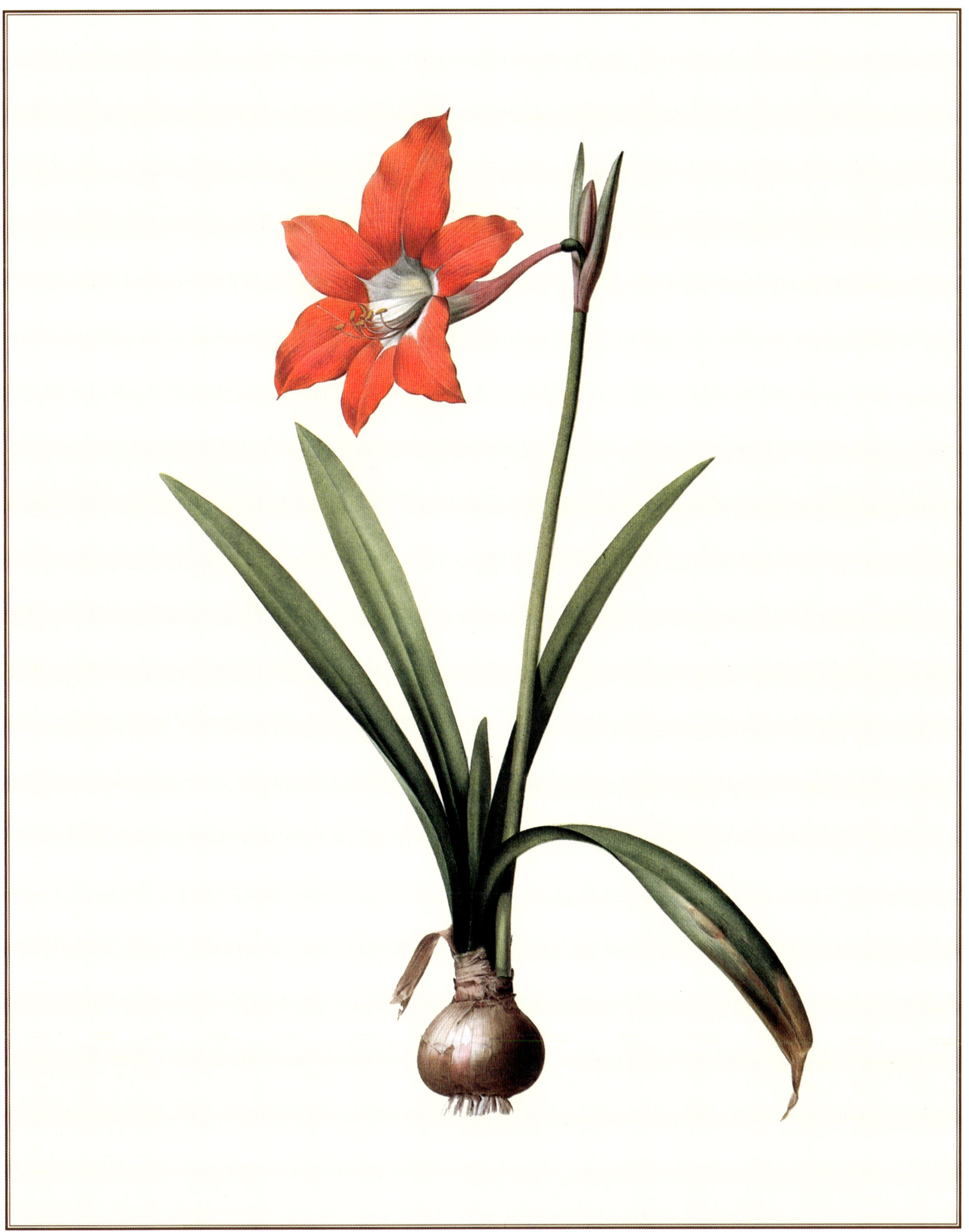

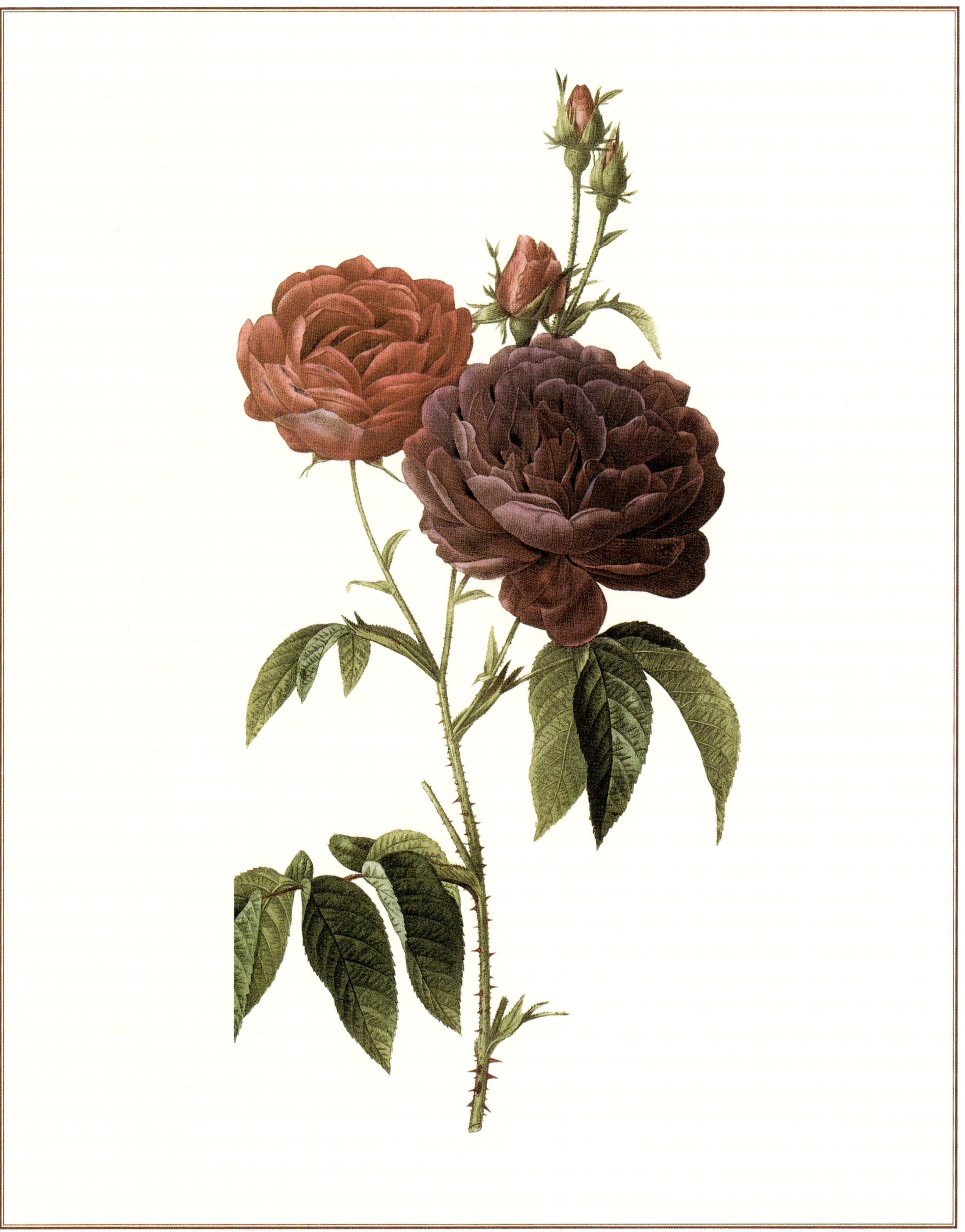

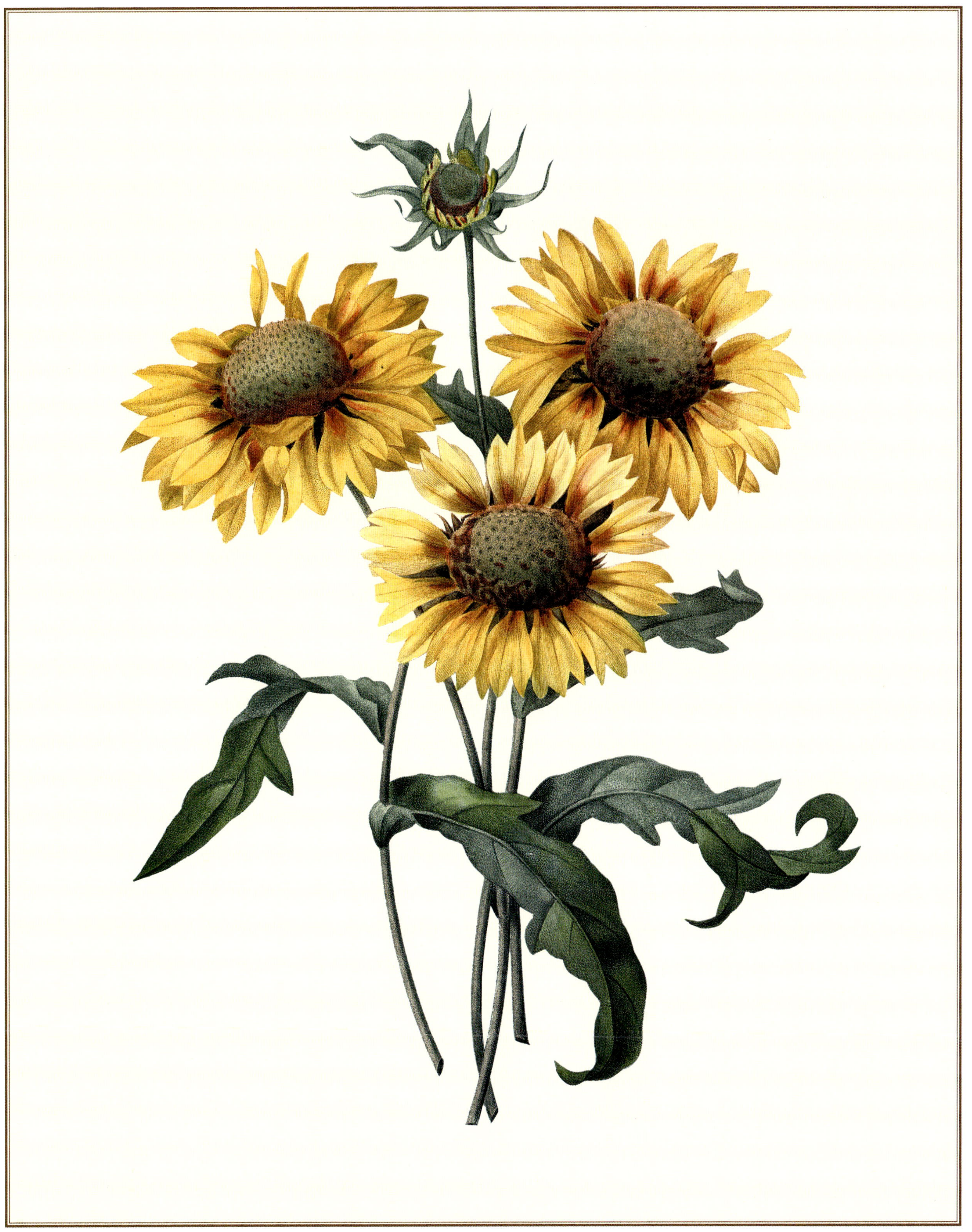

PLATE 10

BLUE PLANTAIN LILY

(Hemerocallis coerulea)

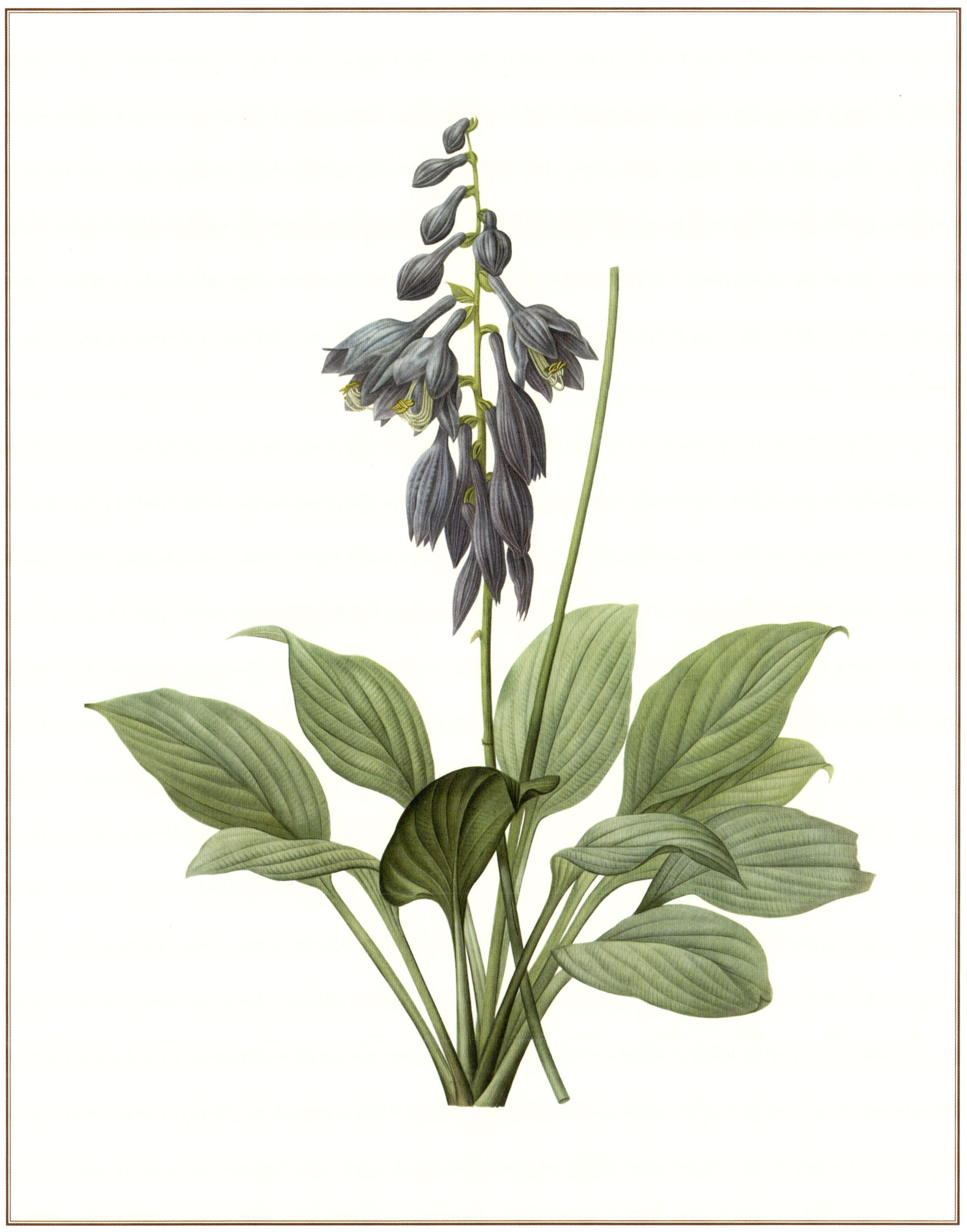

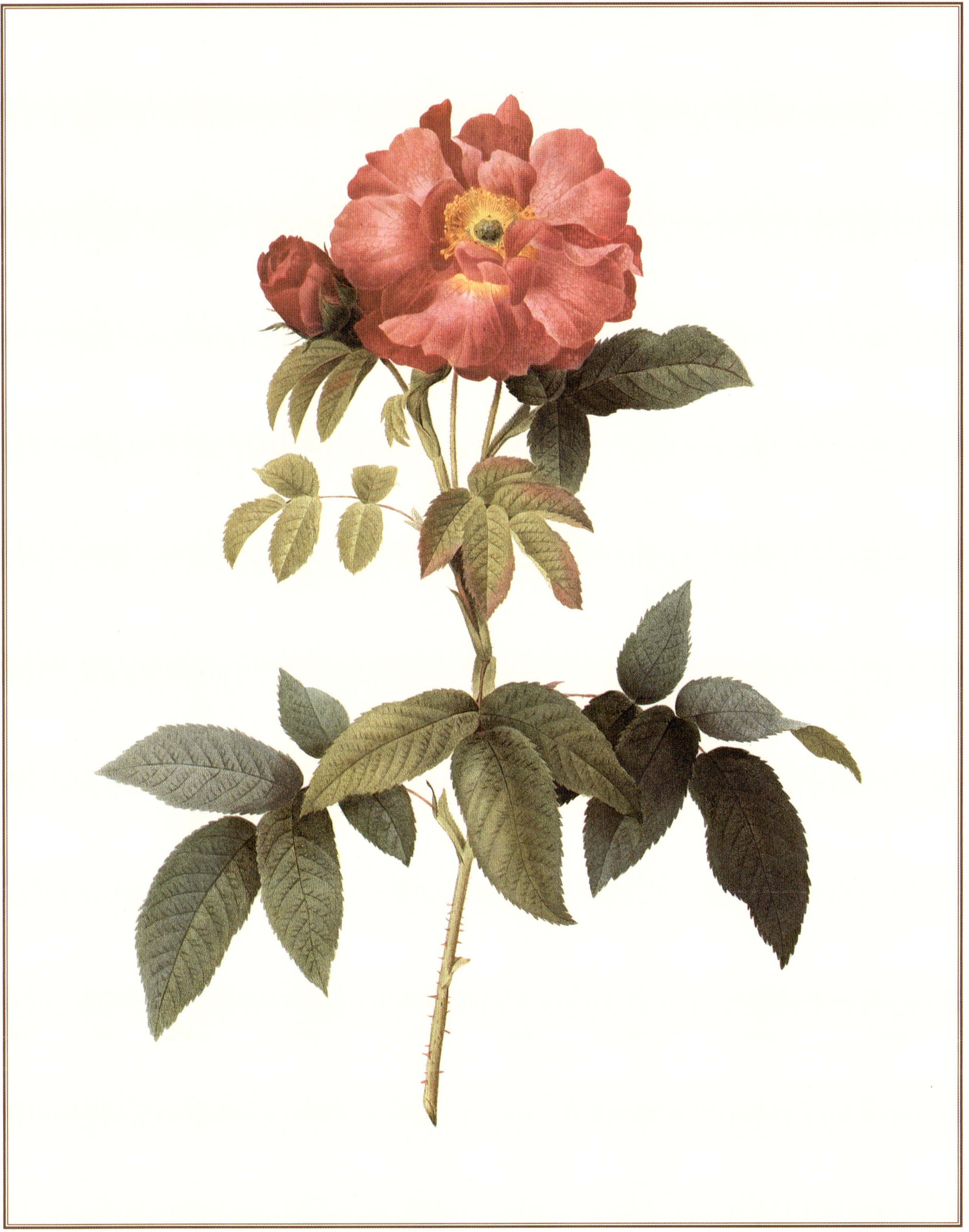

PLATE 12

BOUQUET OF CAMELLIAS, DAFFODILS, AND VIOLETS

(Camellia, Narcissus, and Viola)

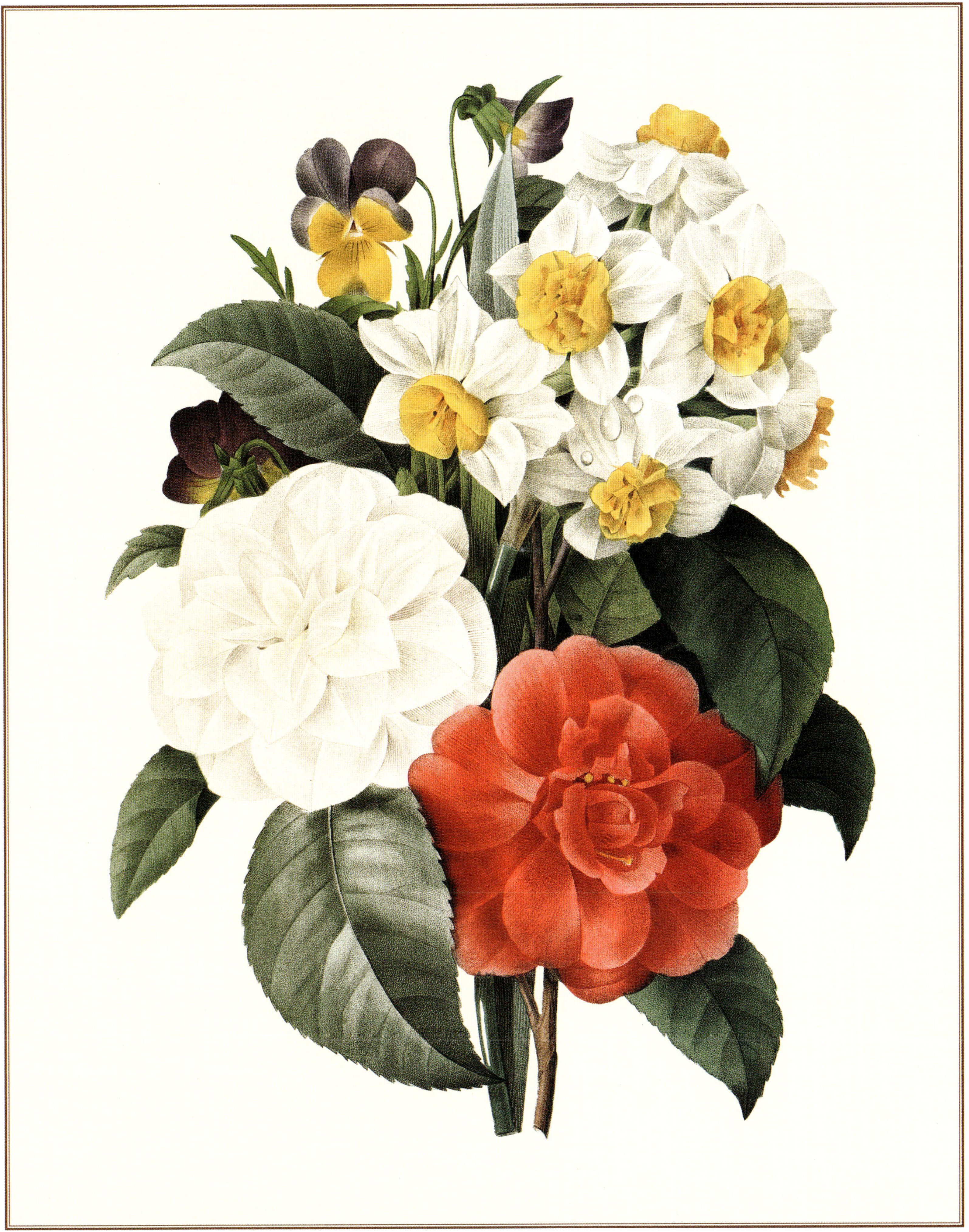

Plate 13
Burnet Rose
(*Rosa pimpinelli marioeburgensis*)

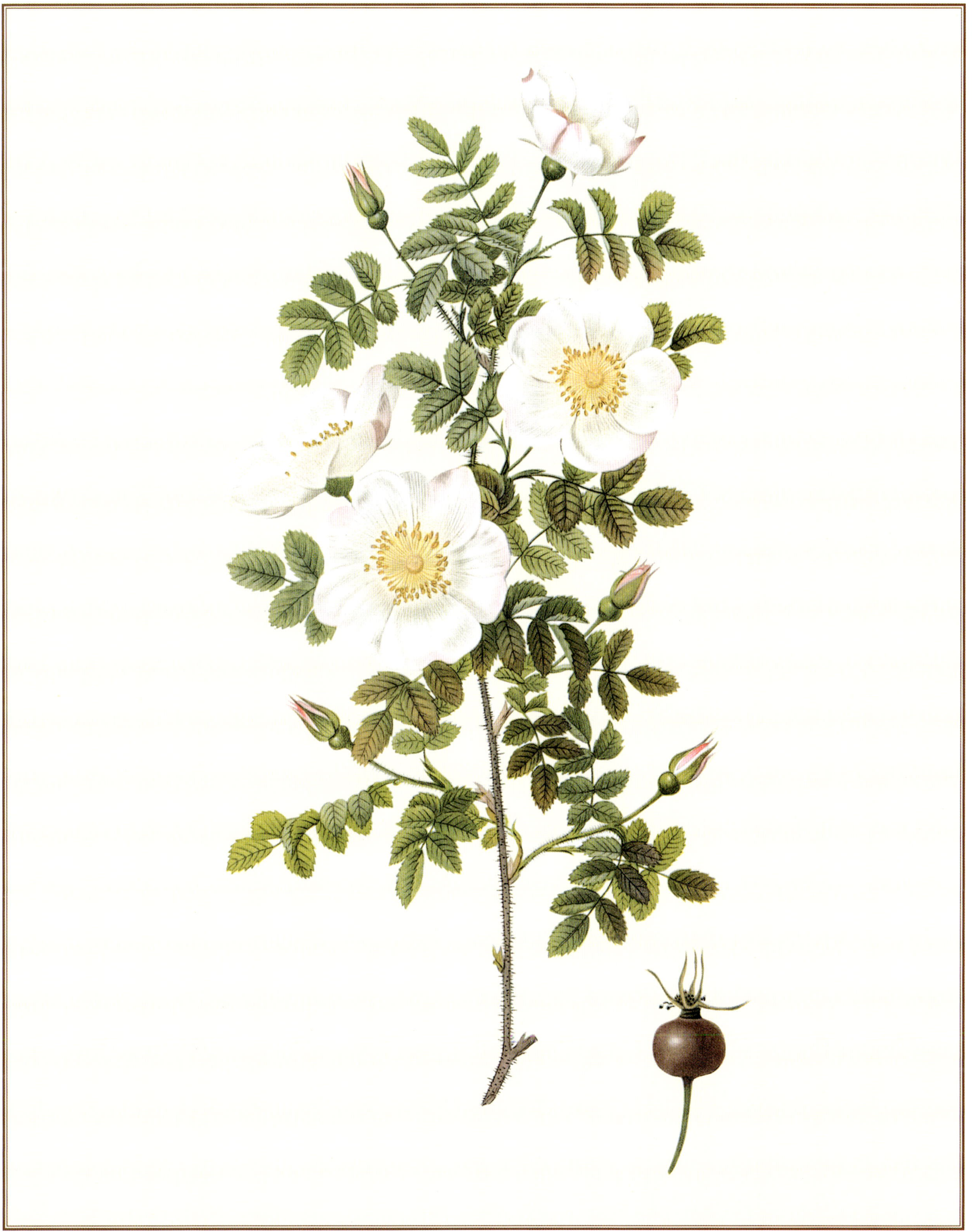

PLATE 14

CABBAGE ROSE

(Rosa centifolia)

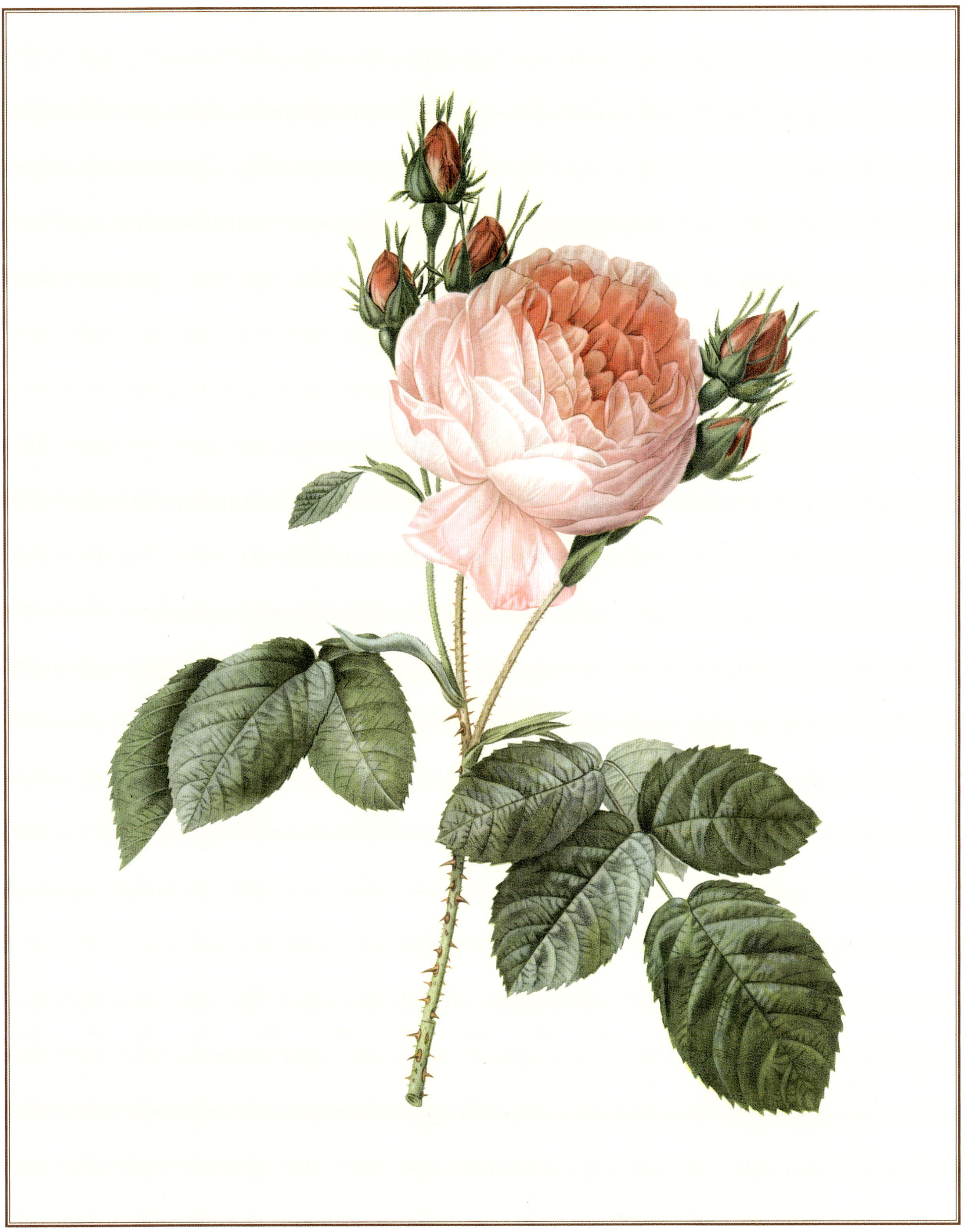

PLATE 15
CANADIAN LILY
(*Lilium penduliflorum*)

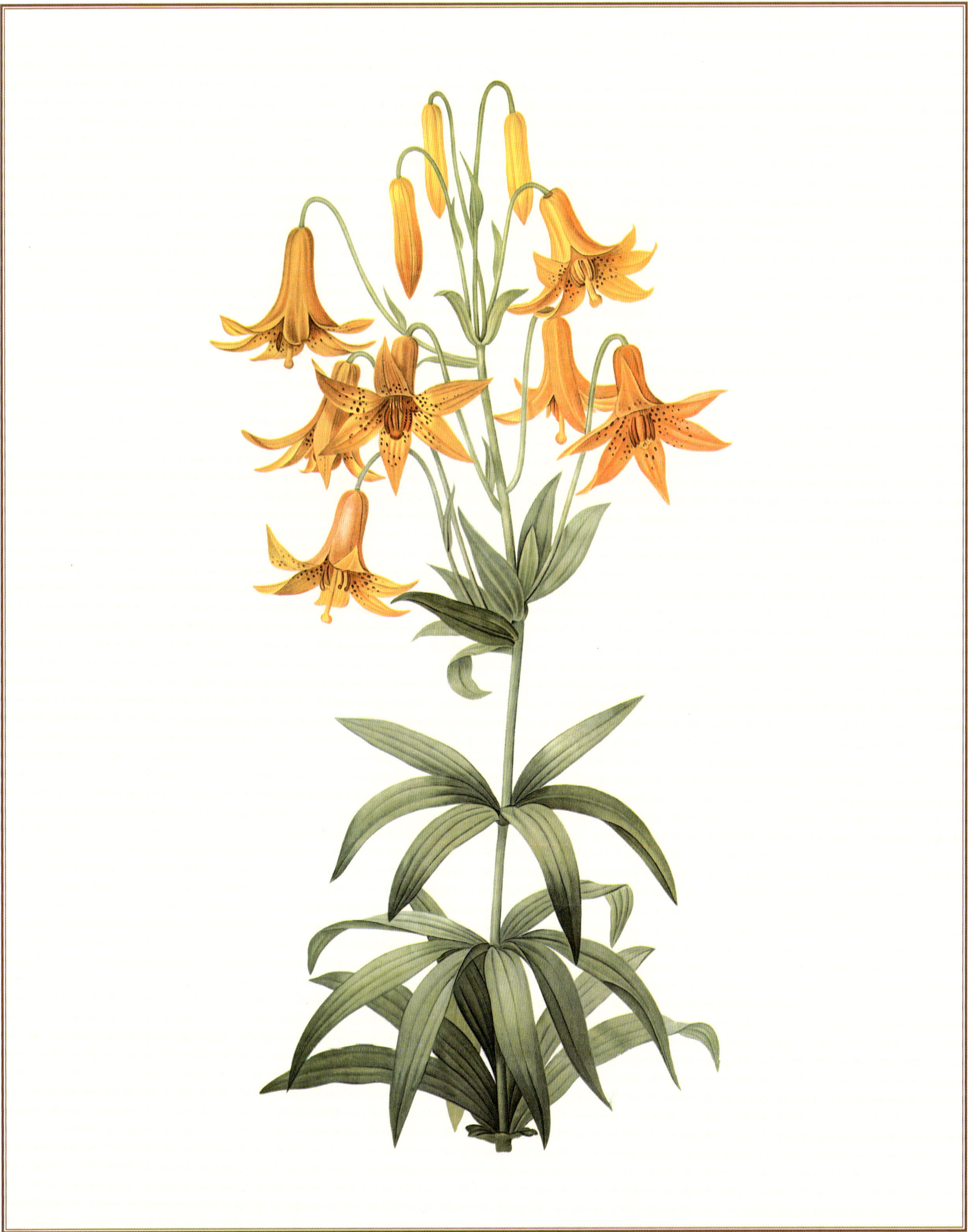

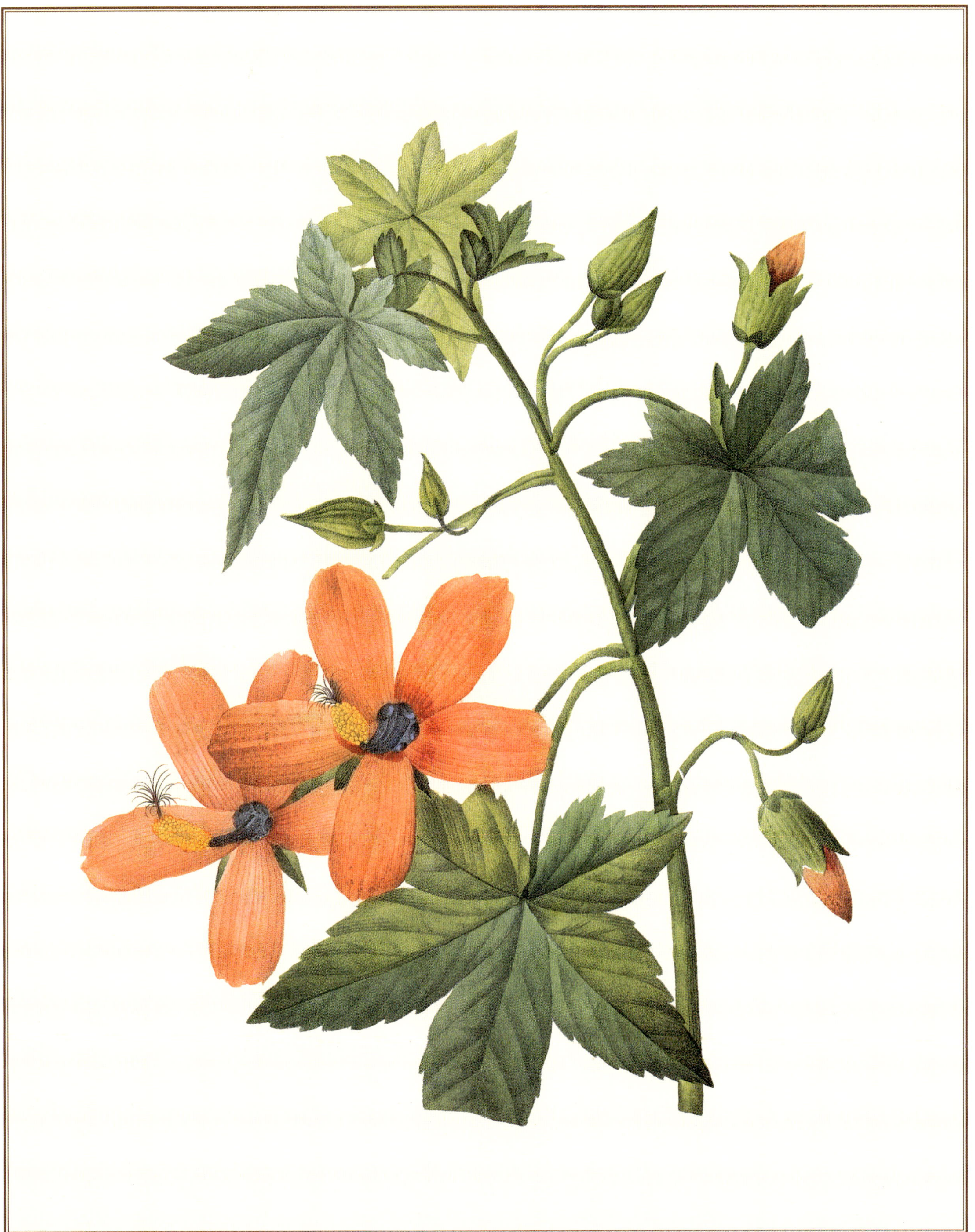

PLATE 17
CANARY YELLOW ROSE
(*Rosa eglanteria luteola*)

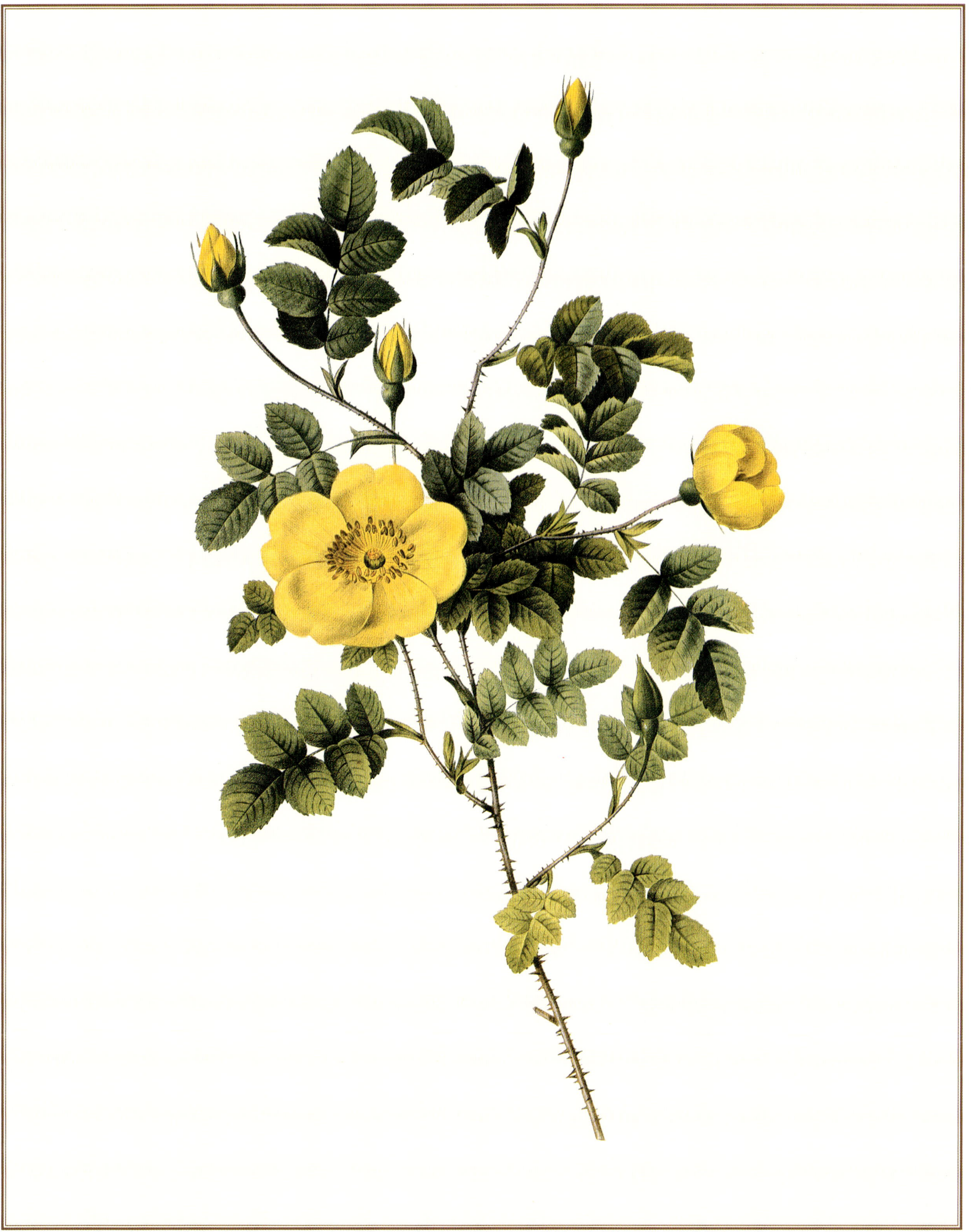

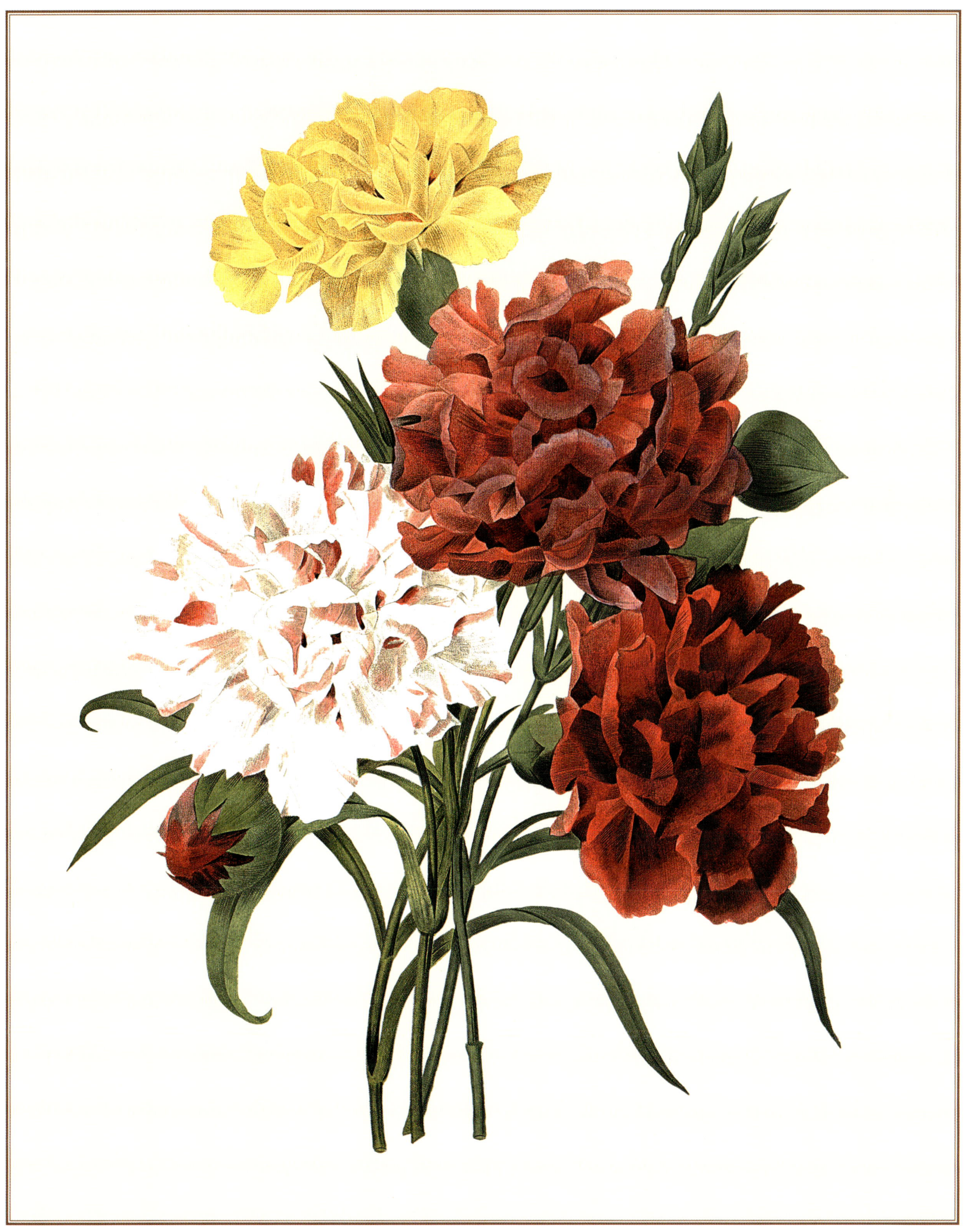

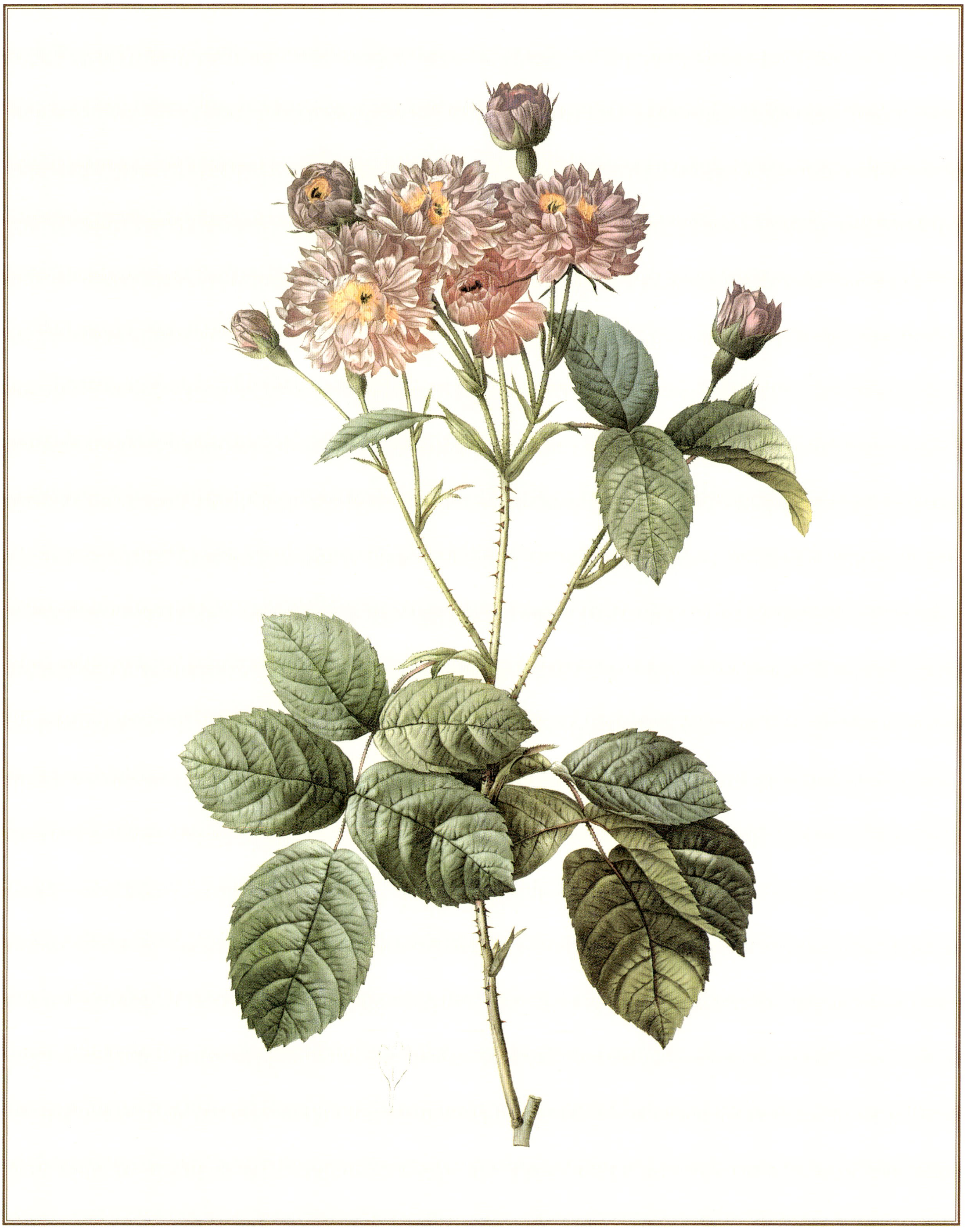

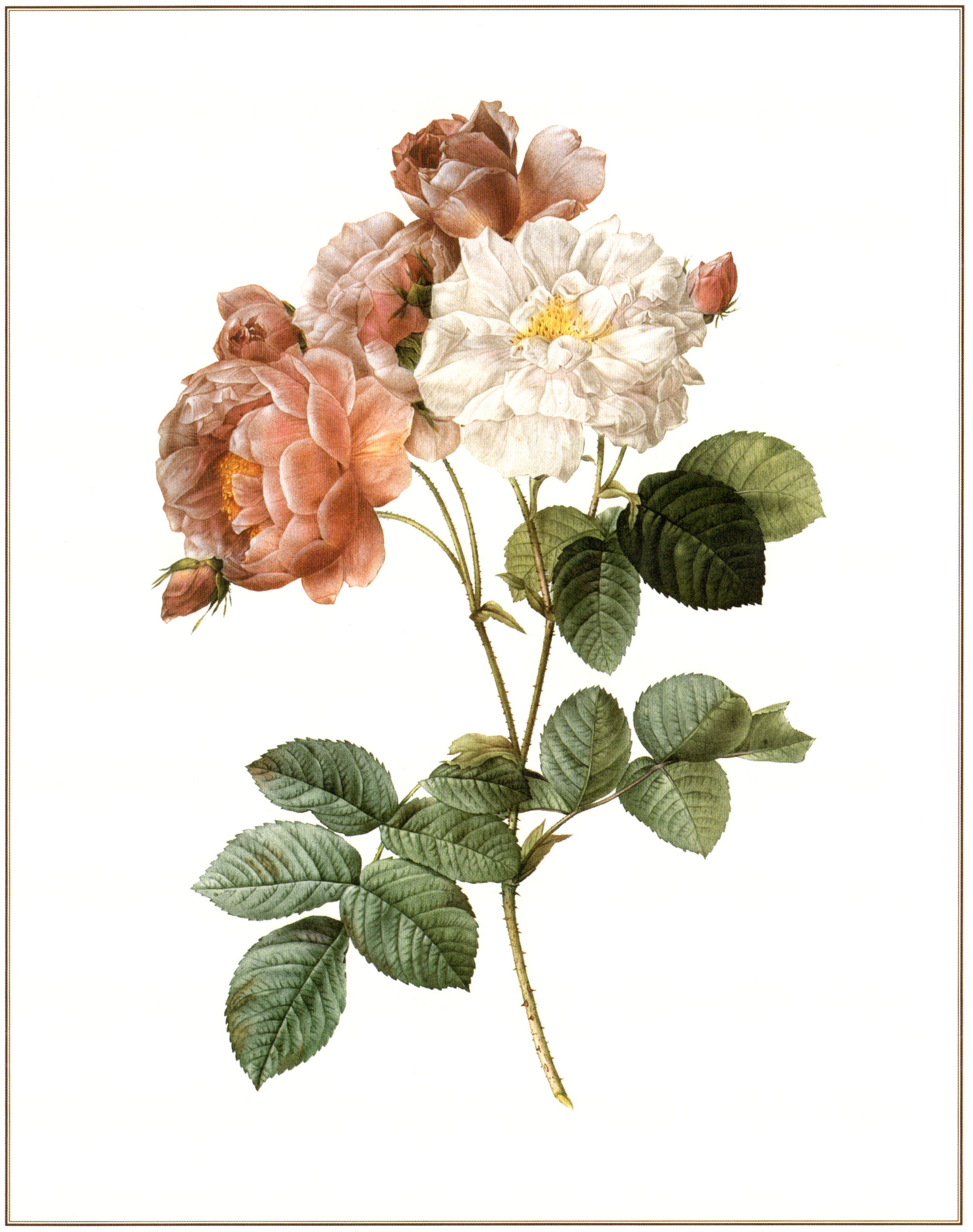

PLATE 21
CHILD OF FRANCE
(*Rosa gallica agatha delphiniana*)

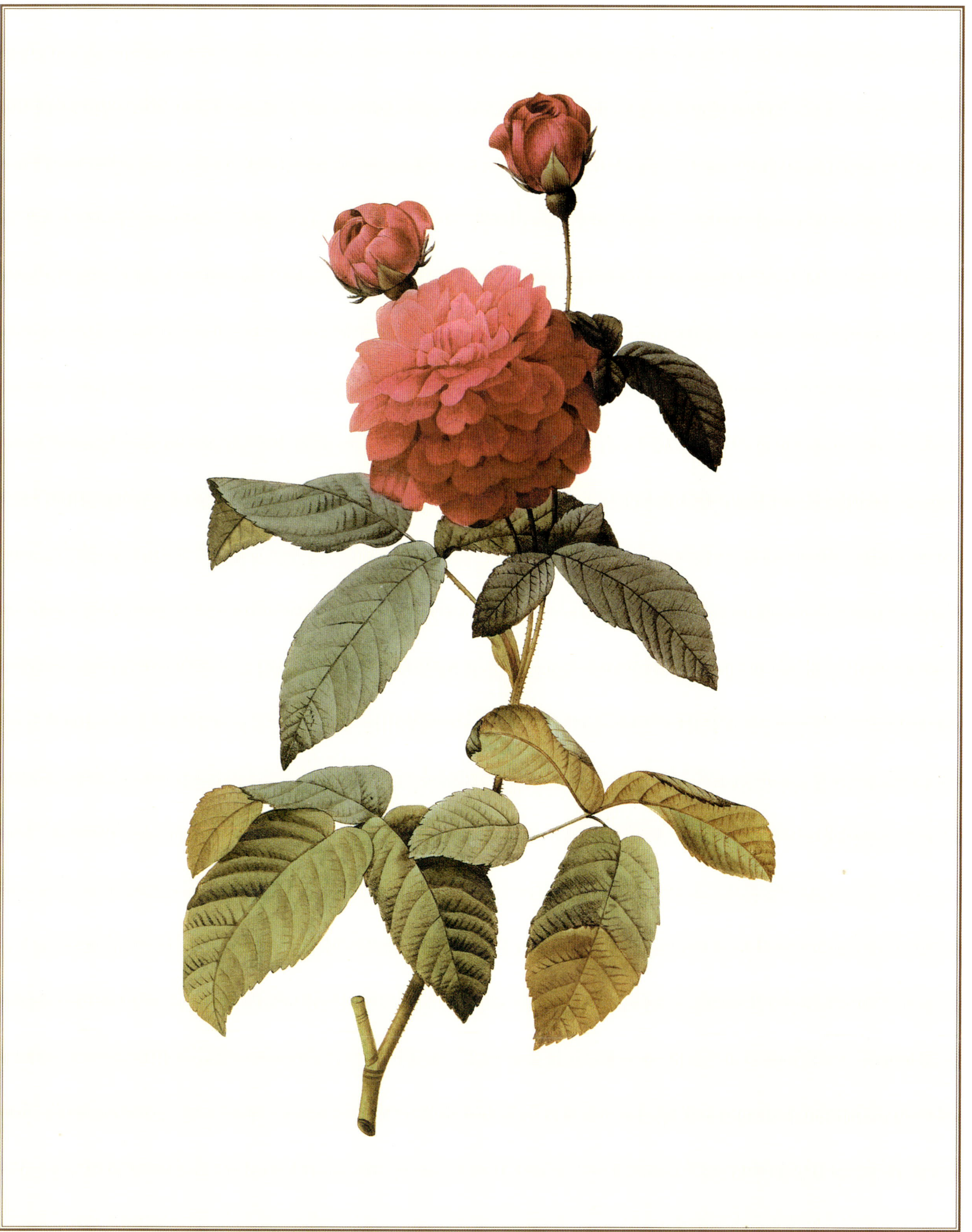

PLATE 22
CHINA ASTER
(*Callistephus chinensis*)

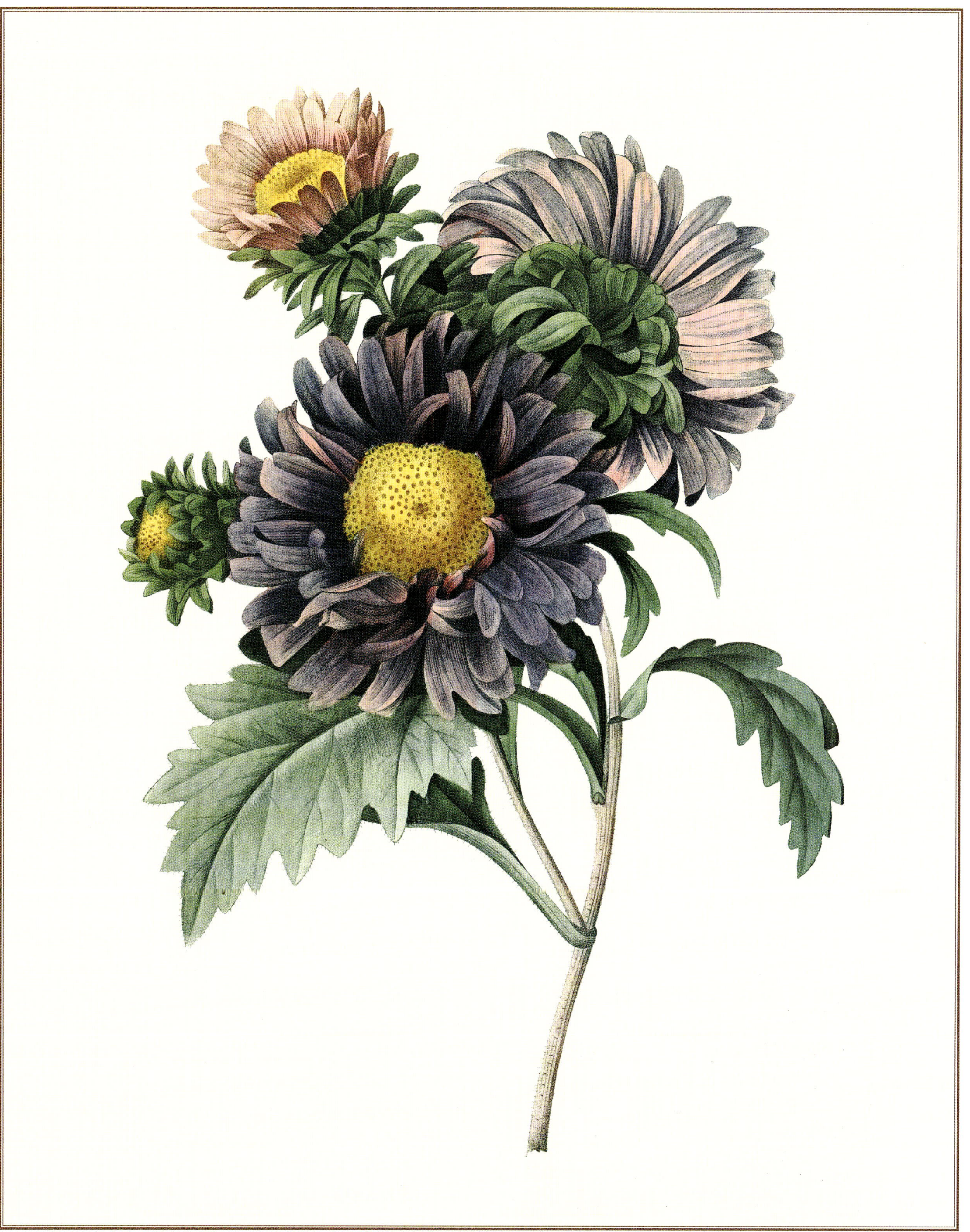

PLATE 23
CHINA ROSE
(Rosa indica)

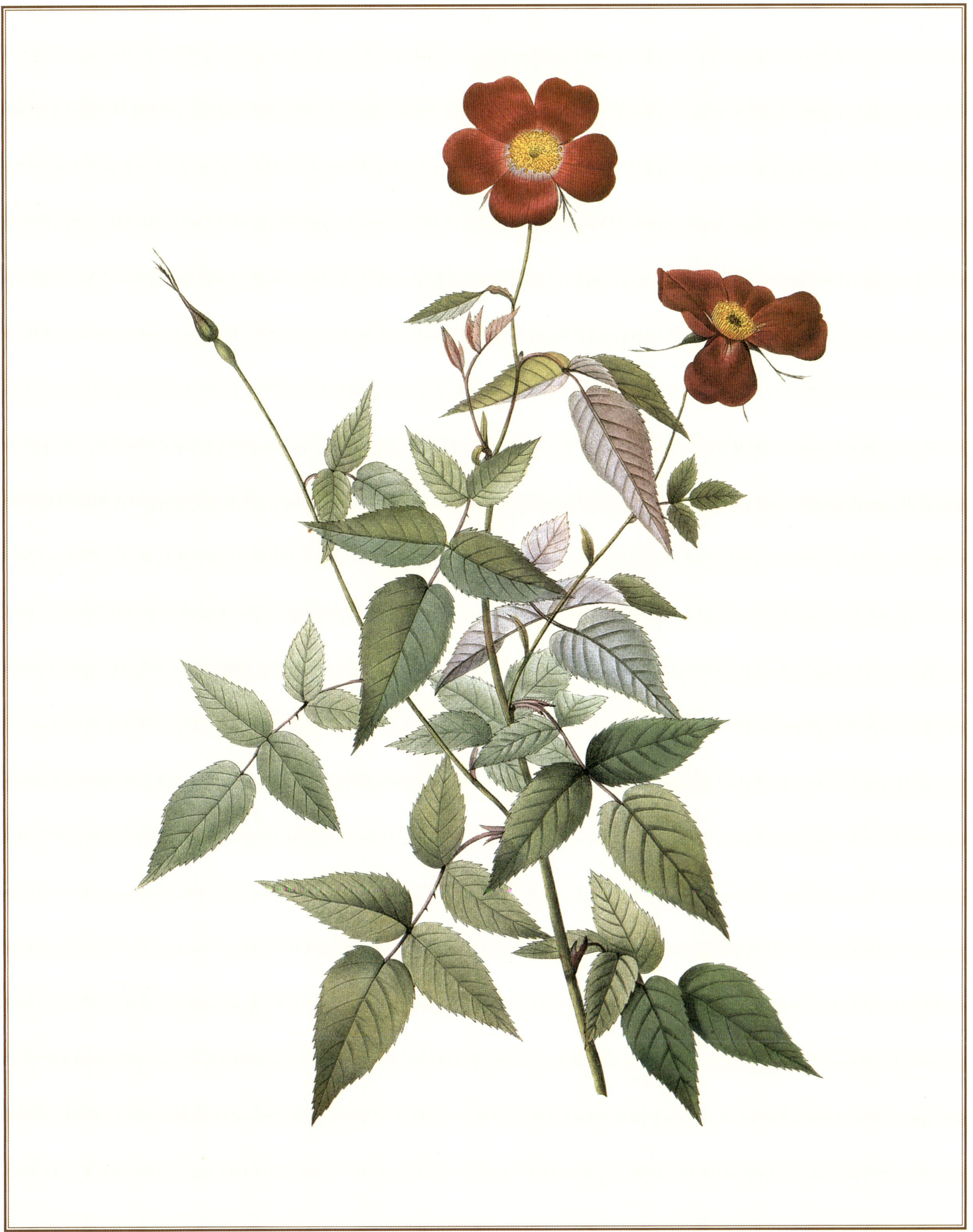

Plate 24

Common Hyacinth

(Hyacinthus orientalis linnaeus)

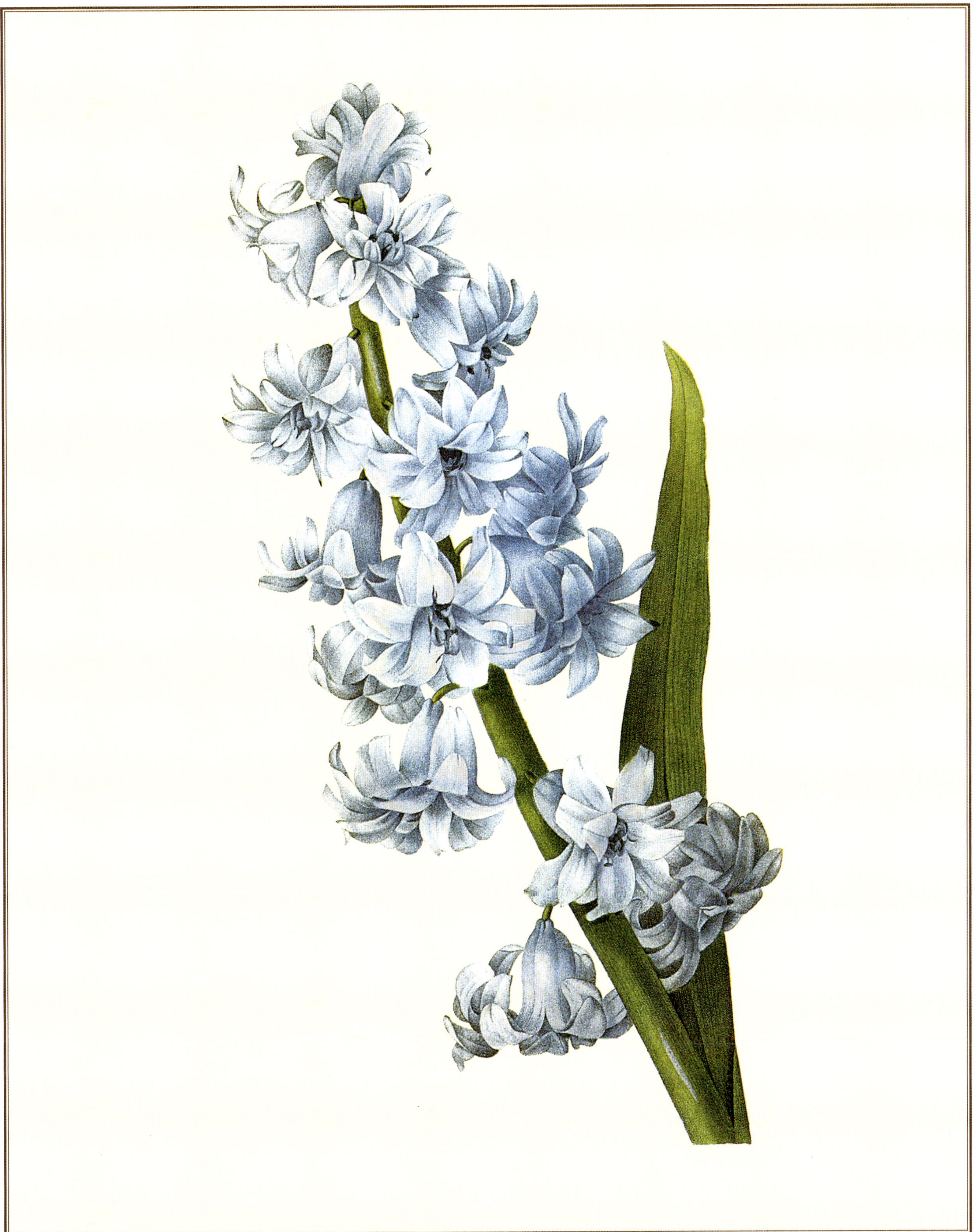

PLATE 25
COMMON PEONY
(*Paeonia officinalis*)

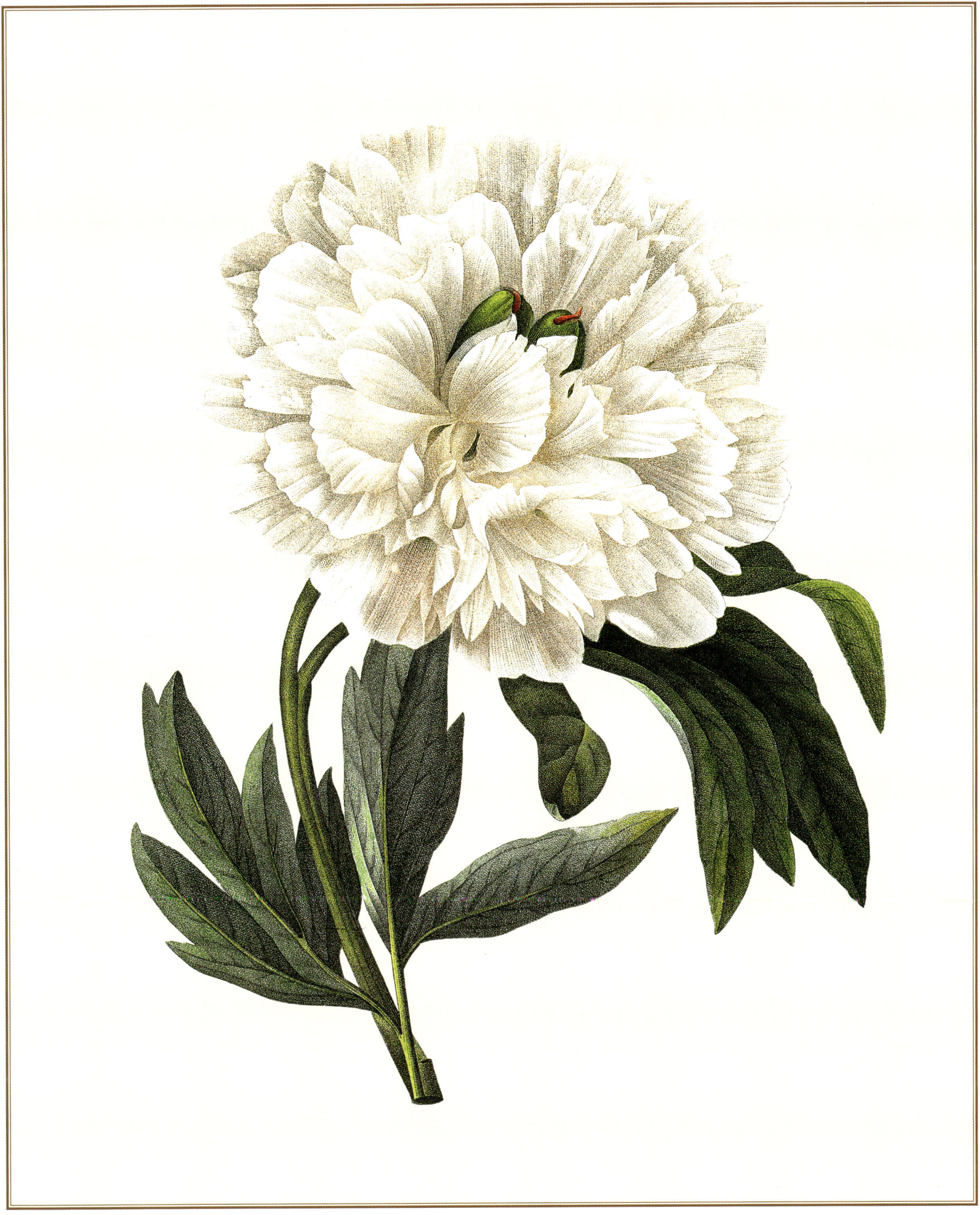

Plate 26
Corymbose Carolina Rose
(*Rosa carolina corymbosa*)

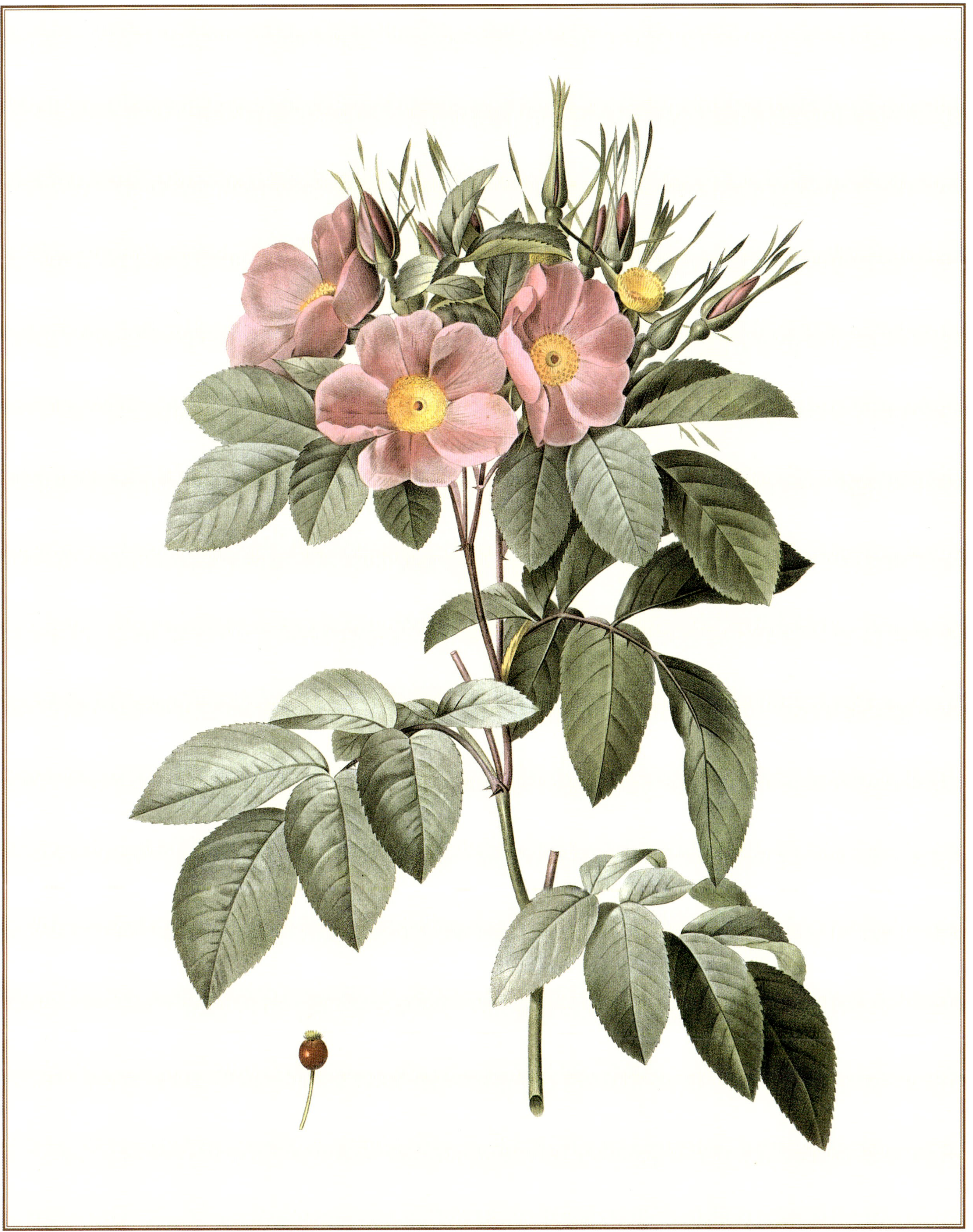

PLATE 27
CROWN IMPERIAL
(*Fritillaria imperialis*)

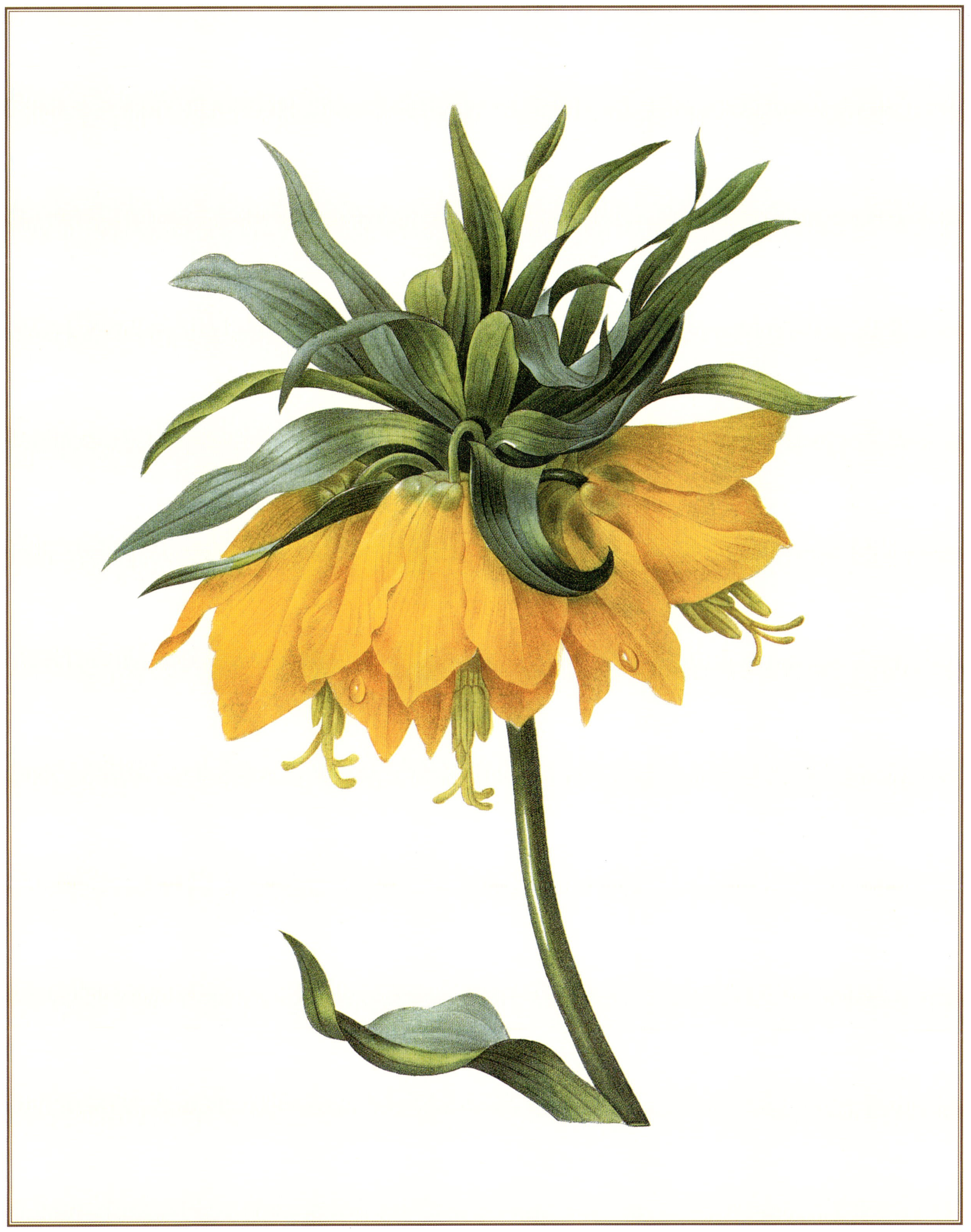

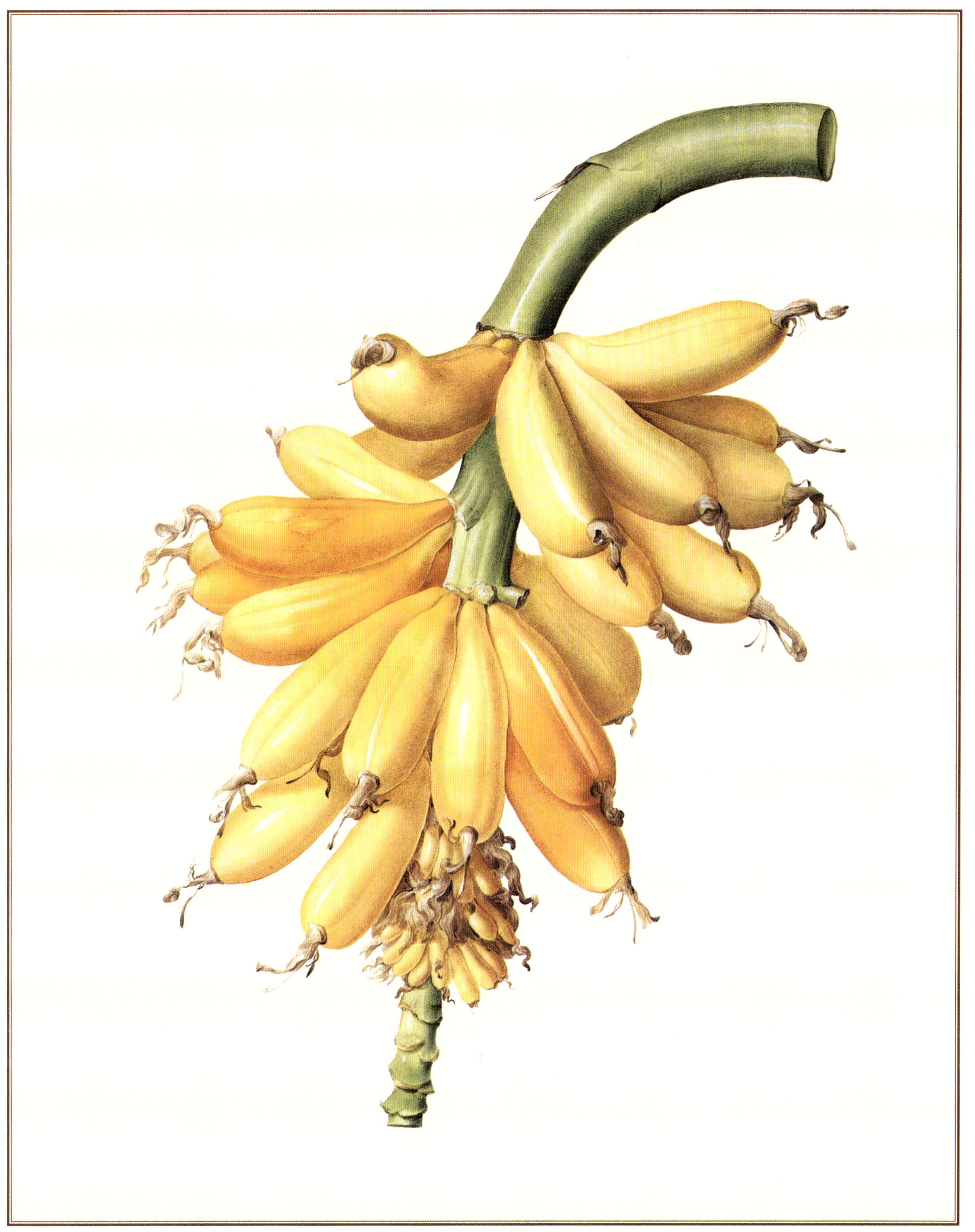

PLATE 29
CUMBERLAND ROSE
(*Rosa centifolia angelica rubra*)

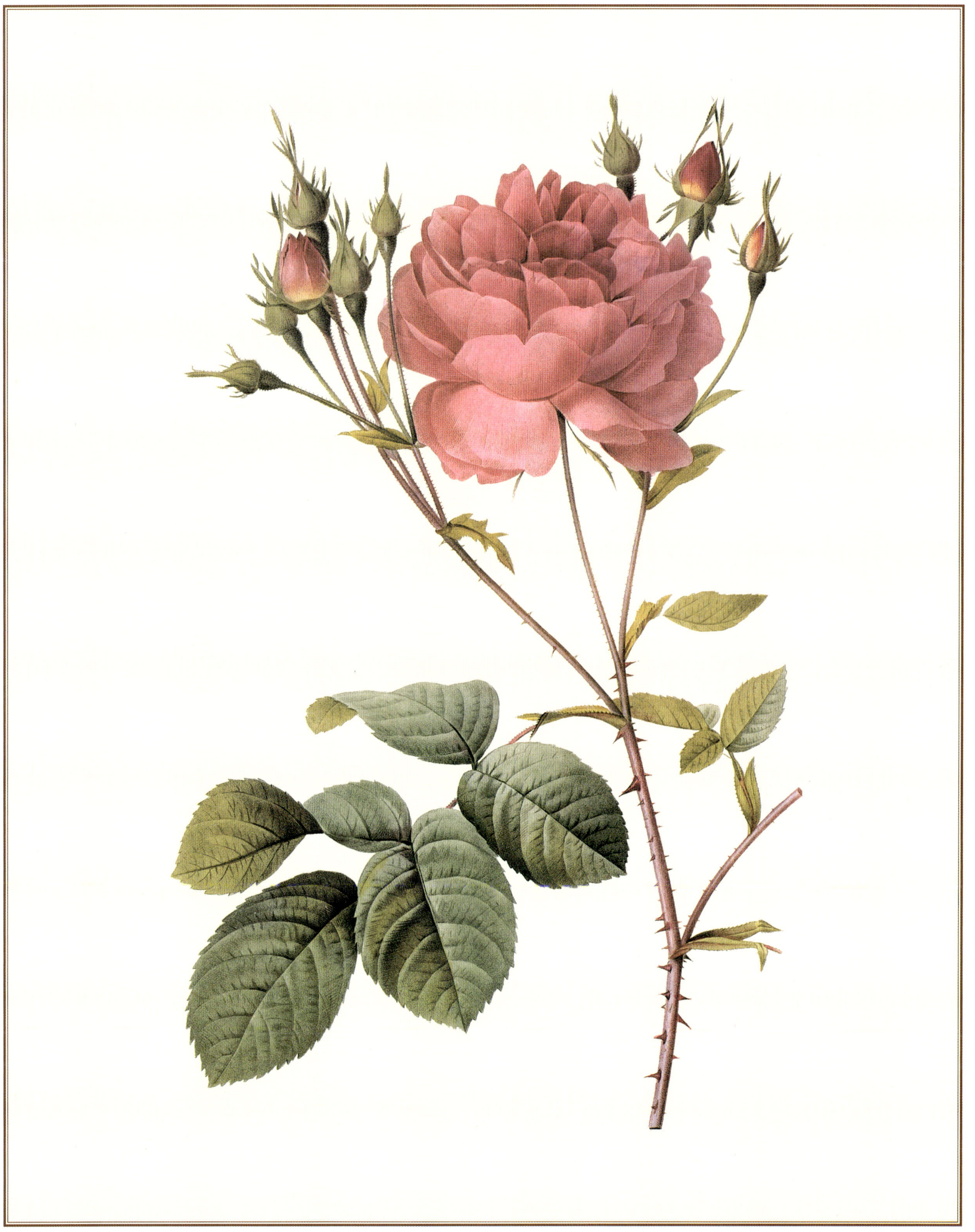

PLATE 30
DALMATIAN IRIS
(*Iris pallida*)

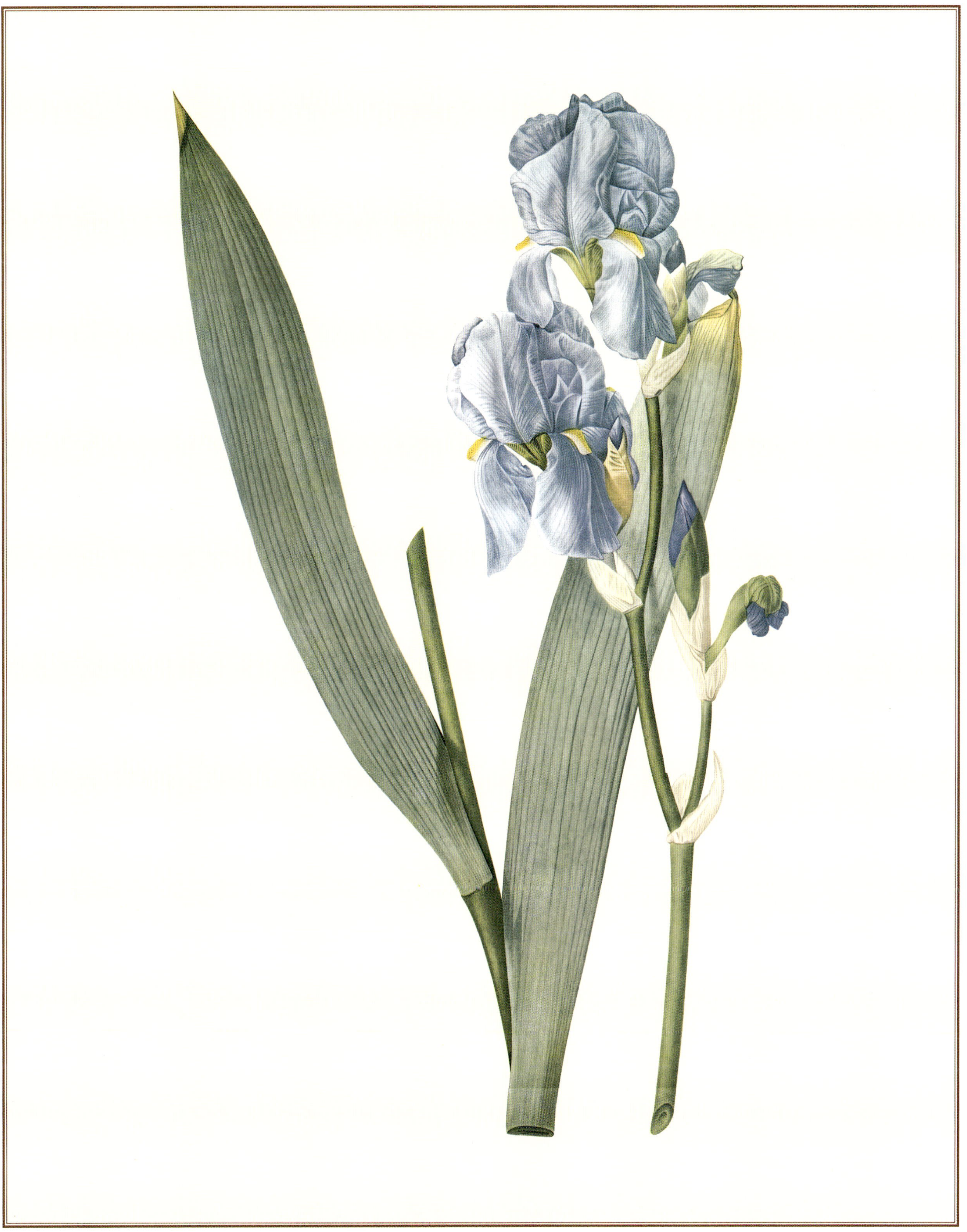

Plate 31

Decumbent Alpine Rose

(*Rosa alpina debilis*)

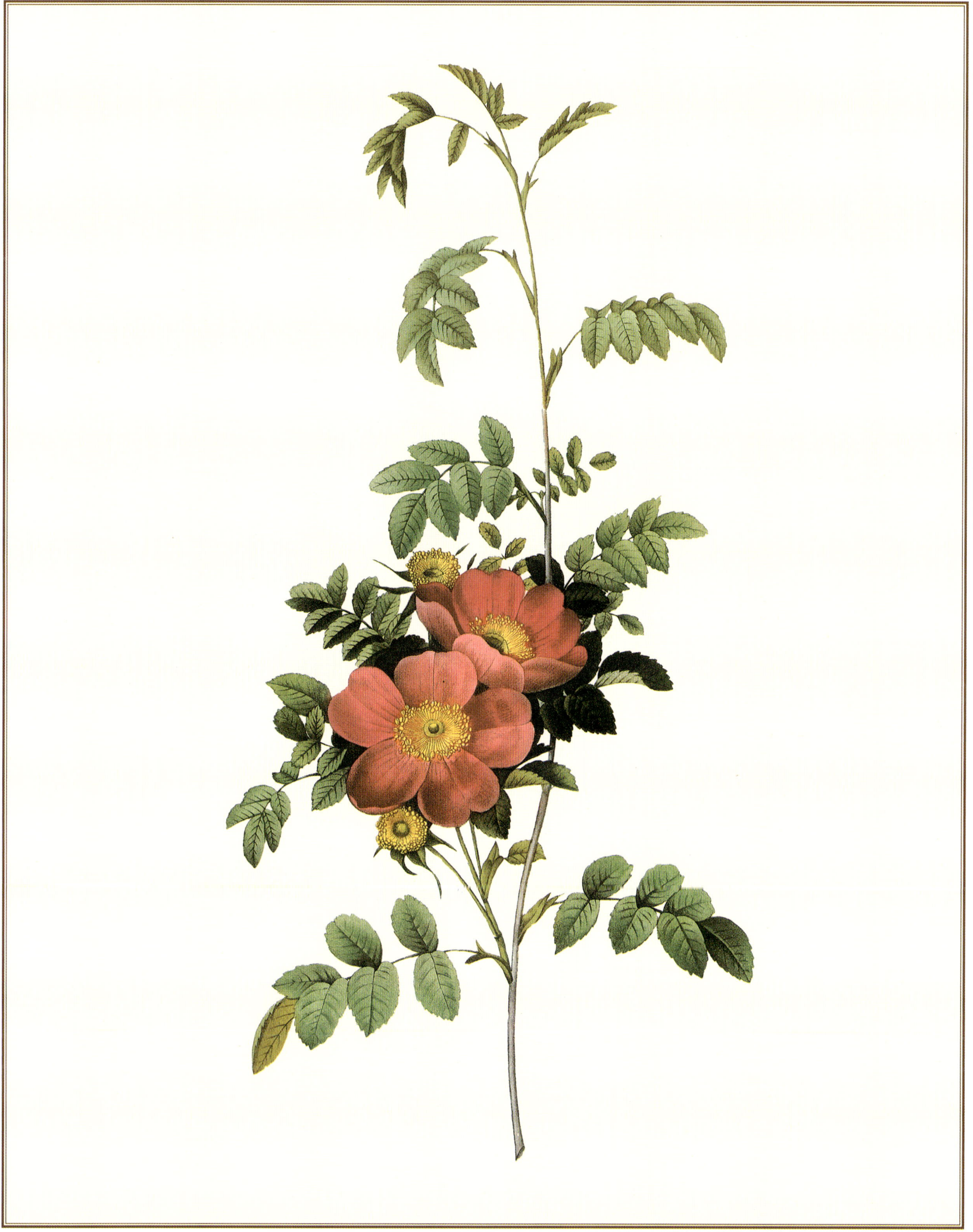

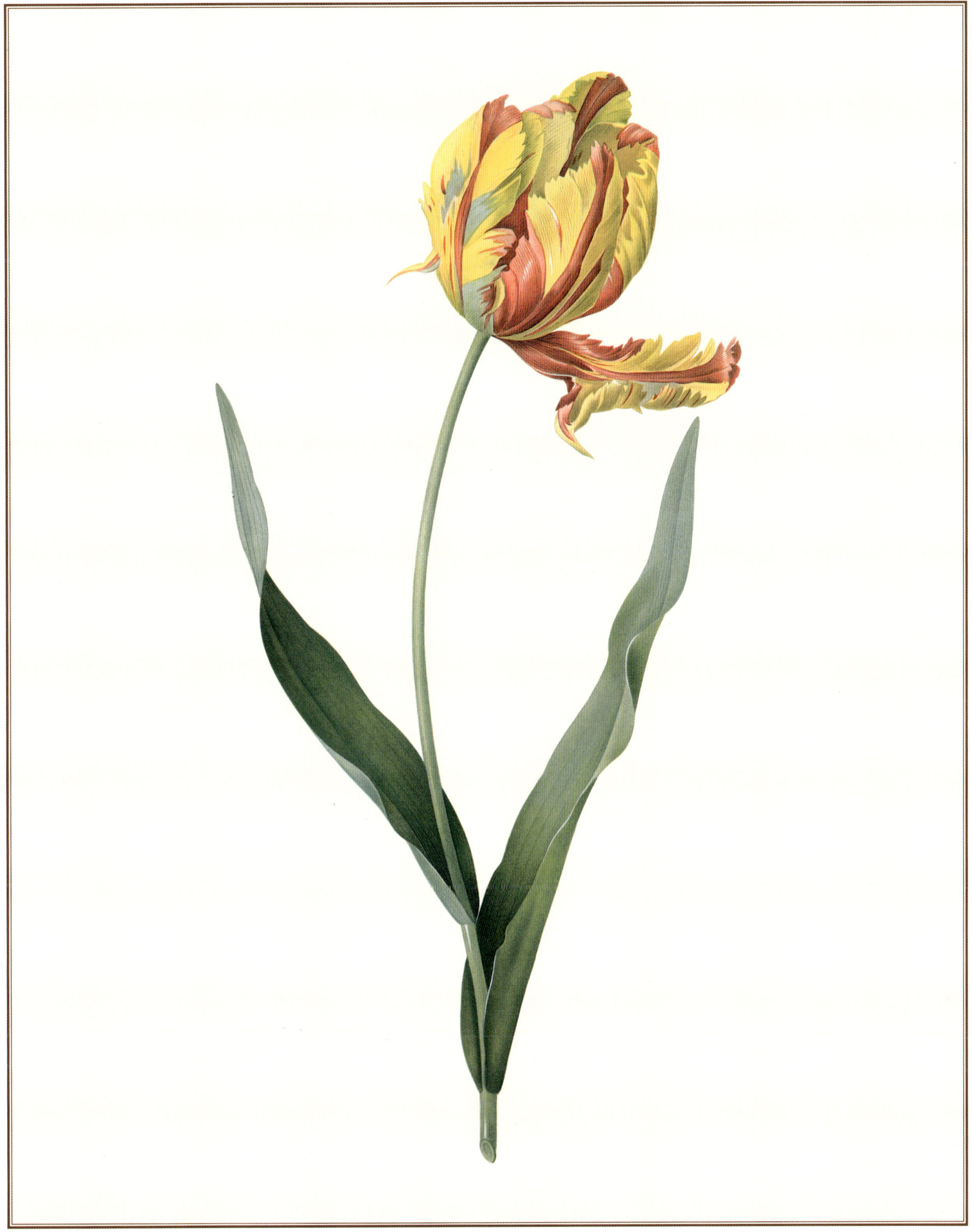

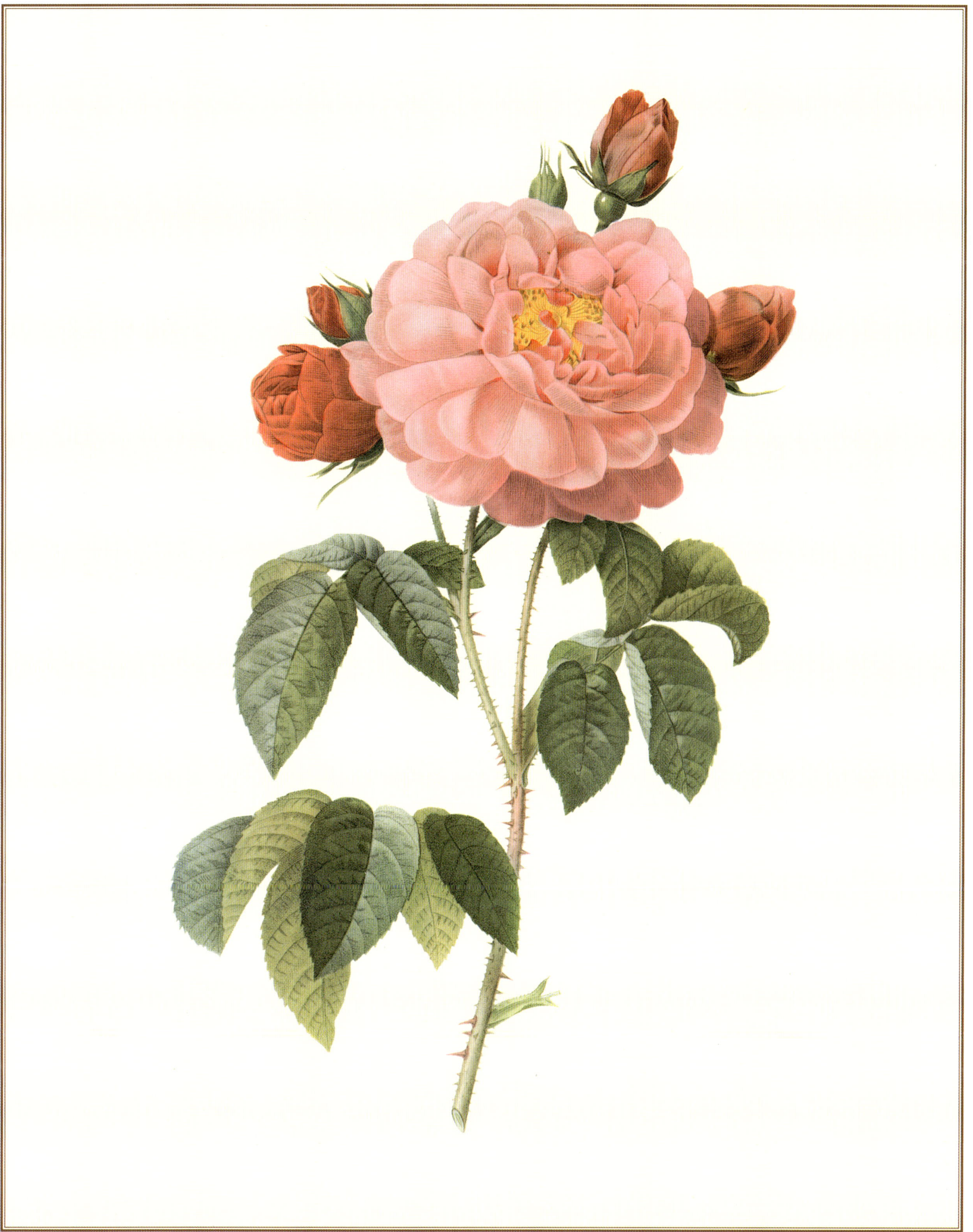

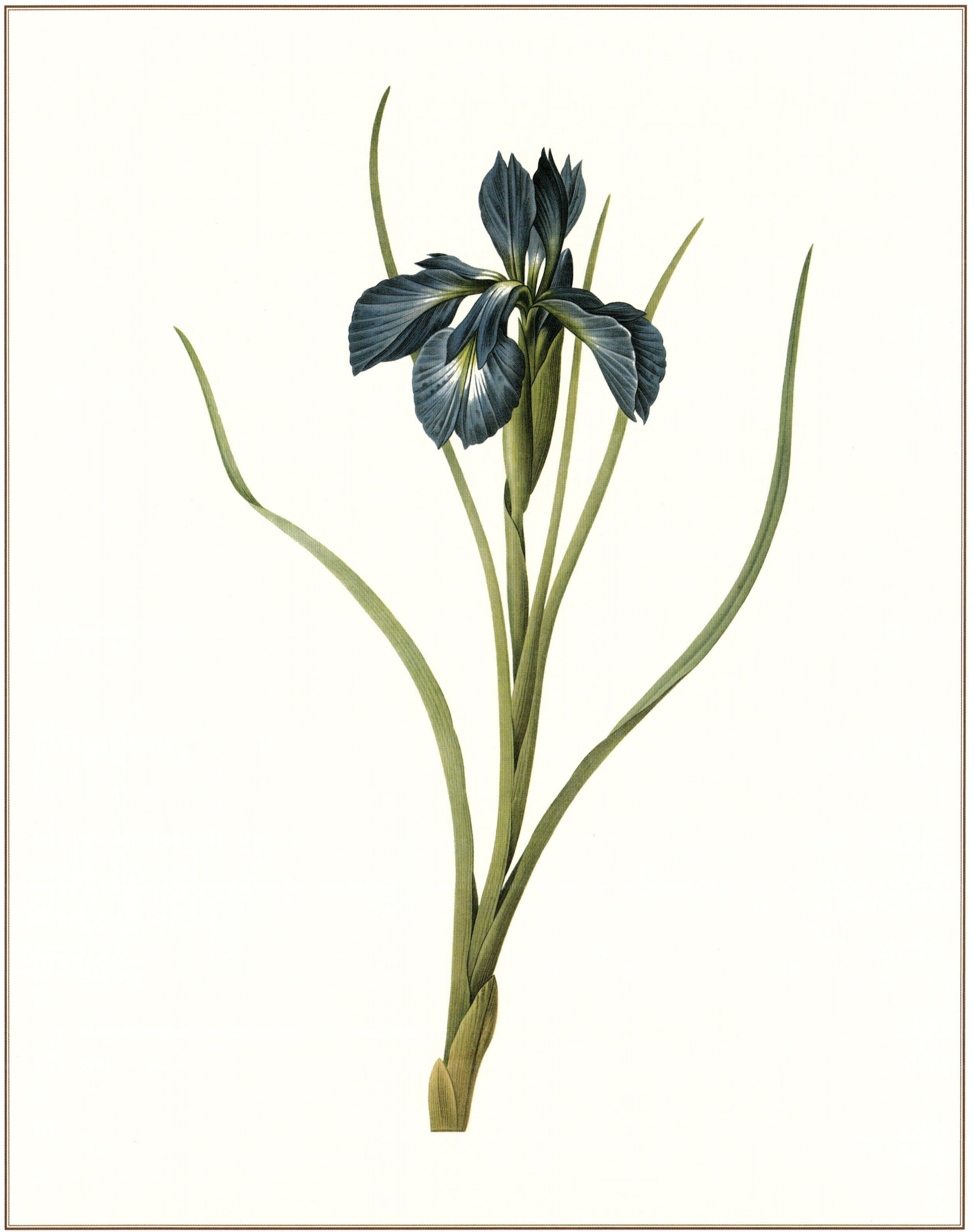

PLATE 35
FRANKFORT ROSE
(*Rosa turbinata*)

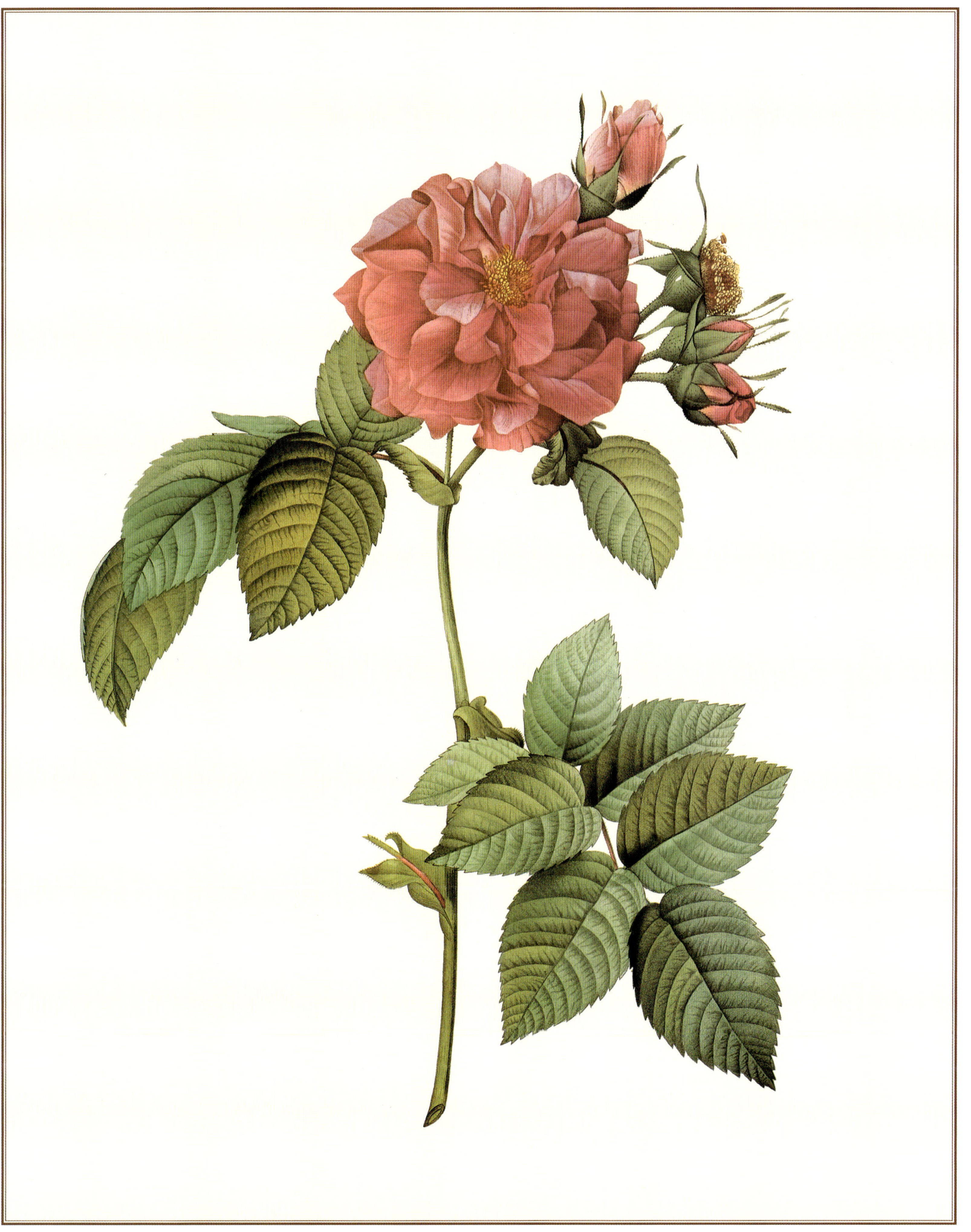

Plate 36
French Marigold
(*Tagetes patula*)

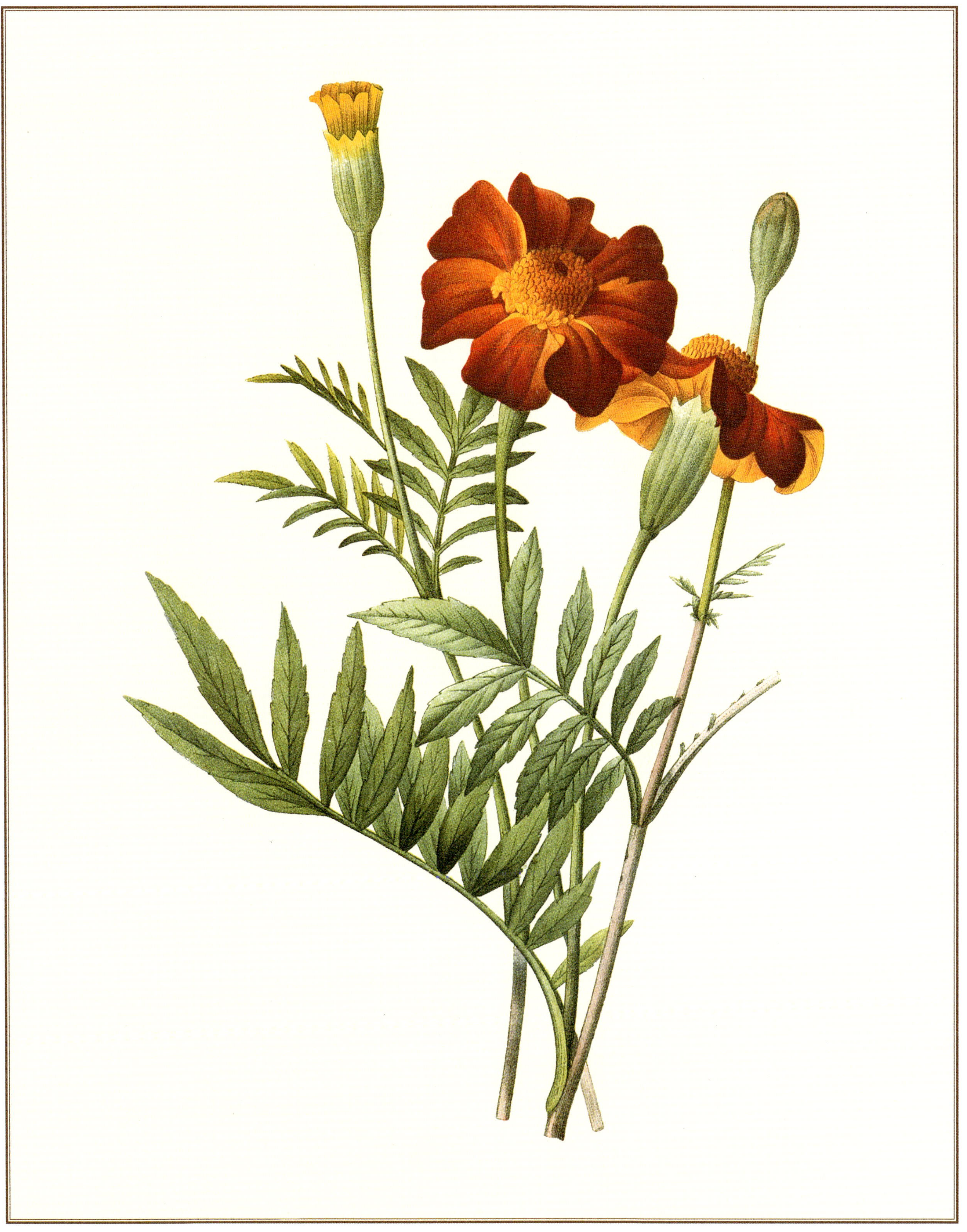

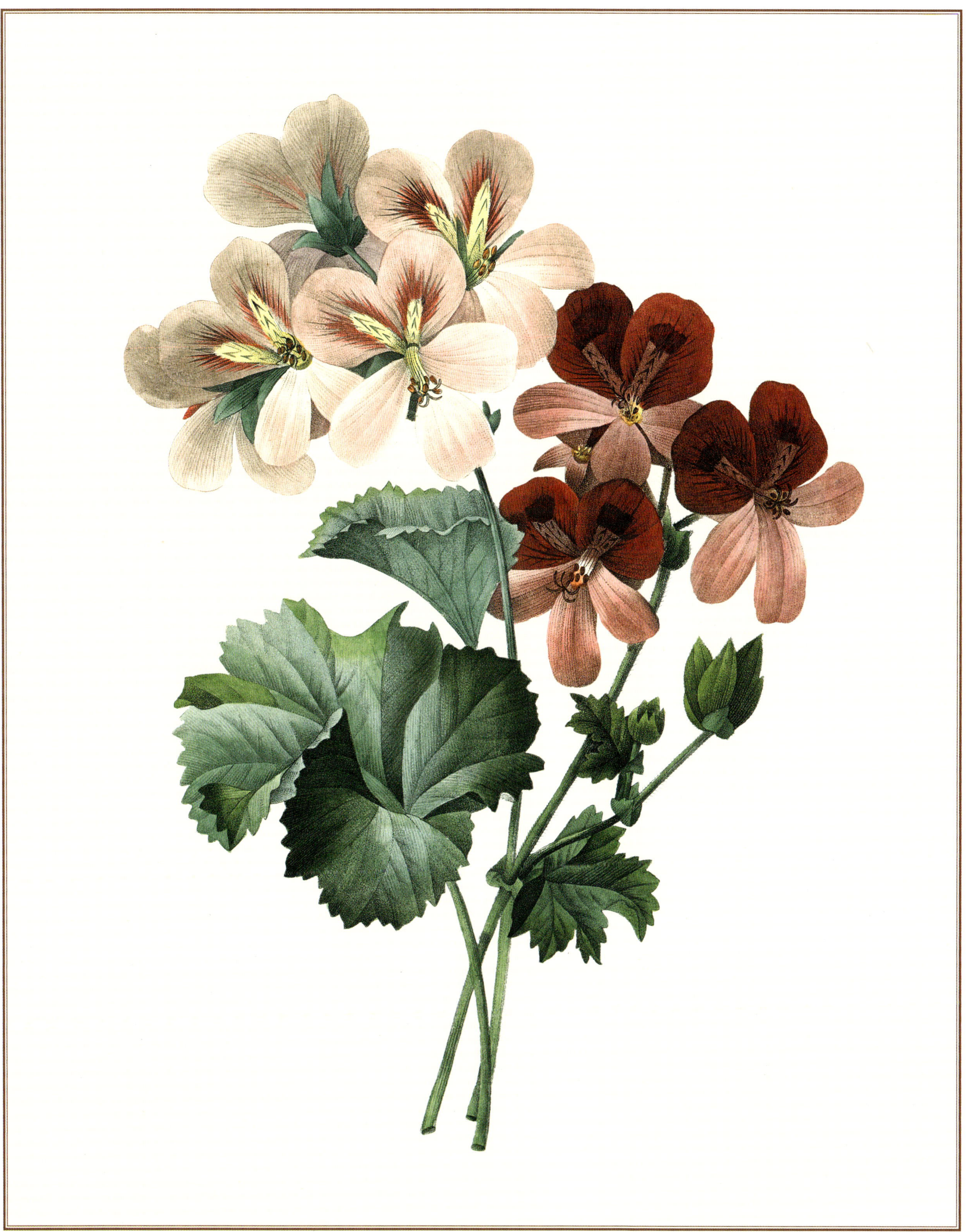

Plate 38

German Flag Iris

(Iris germanica)

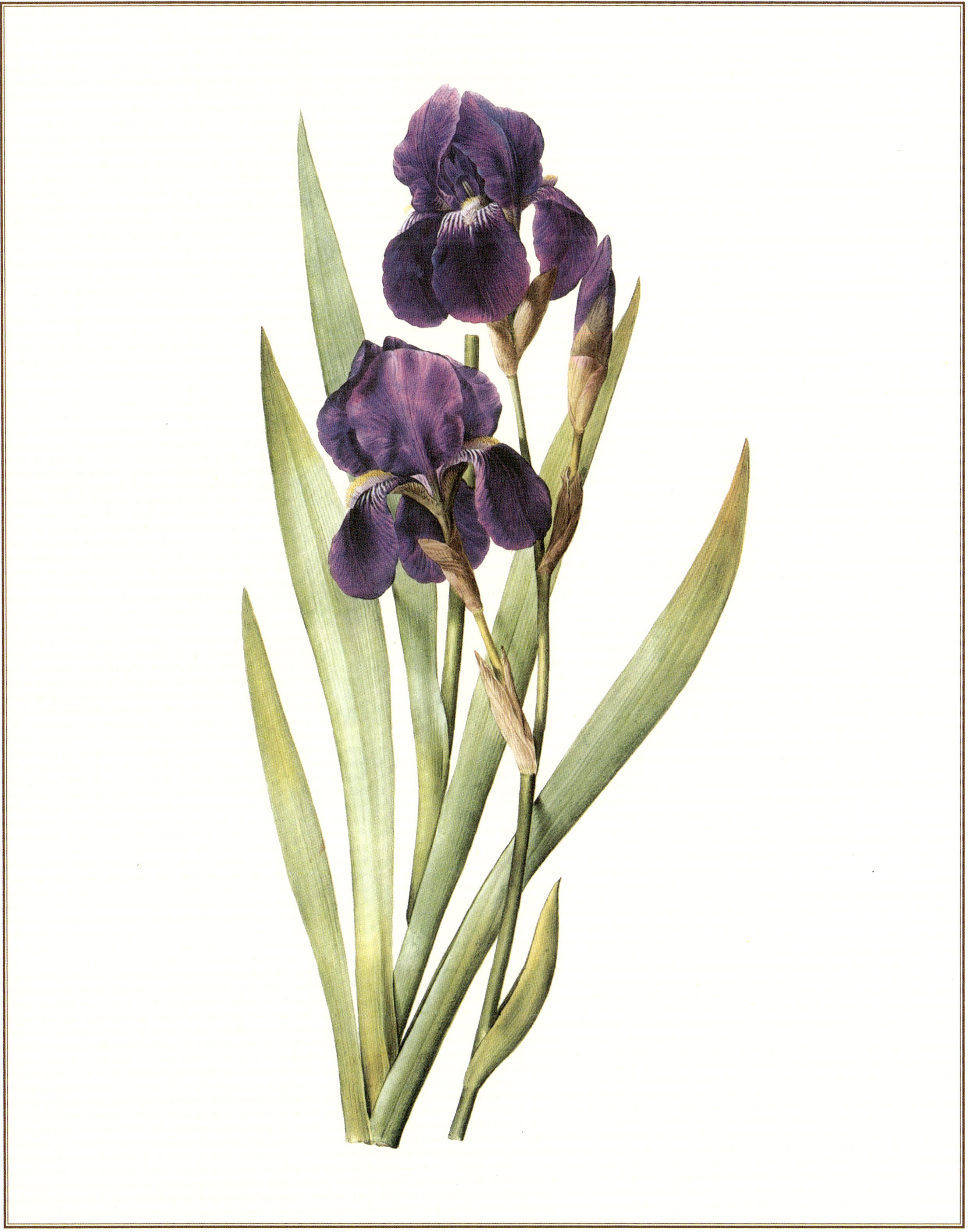

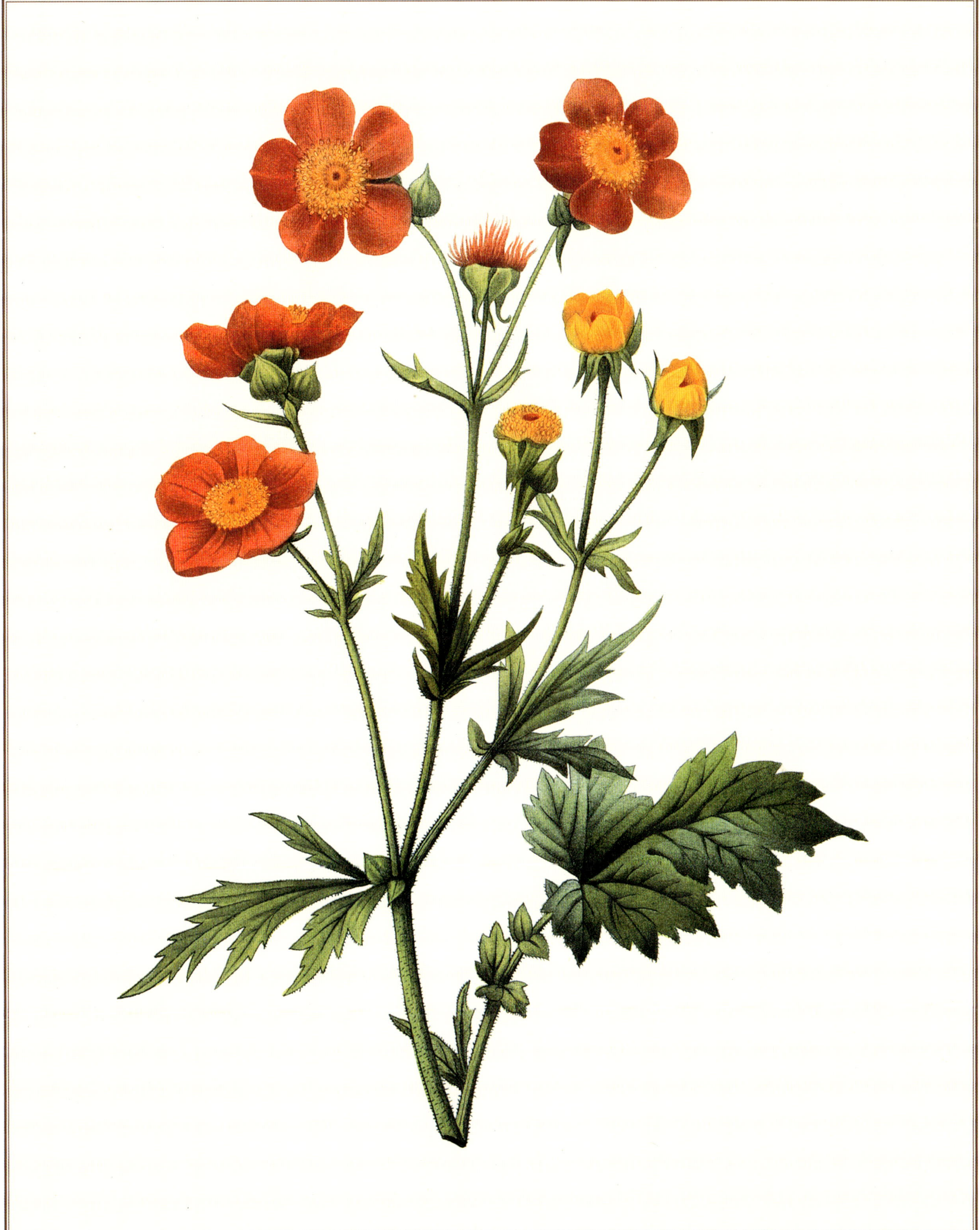

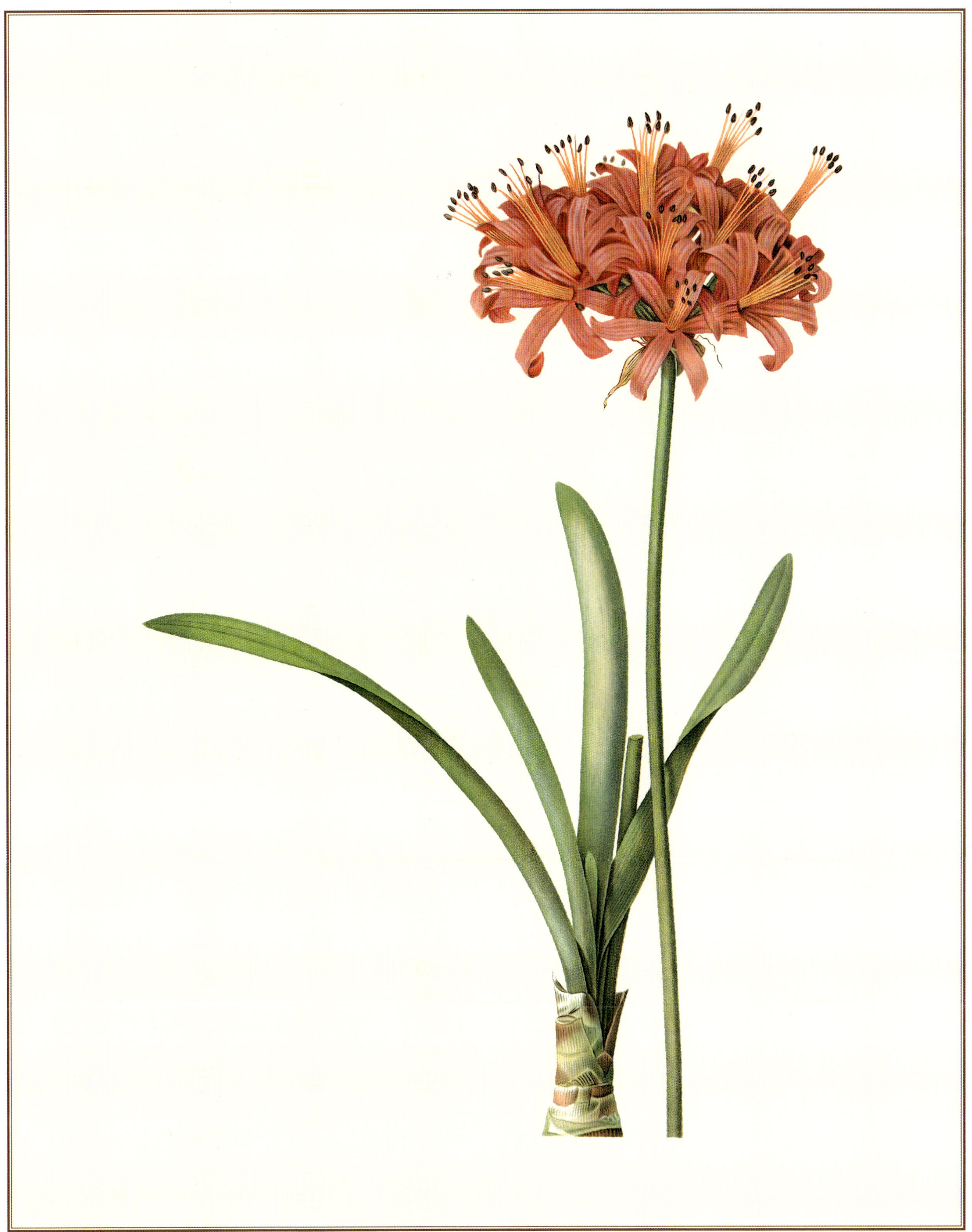

PLATE 41
GLOSSY ROSE
(Rosa lucida)

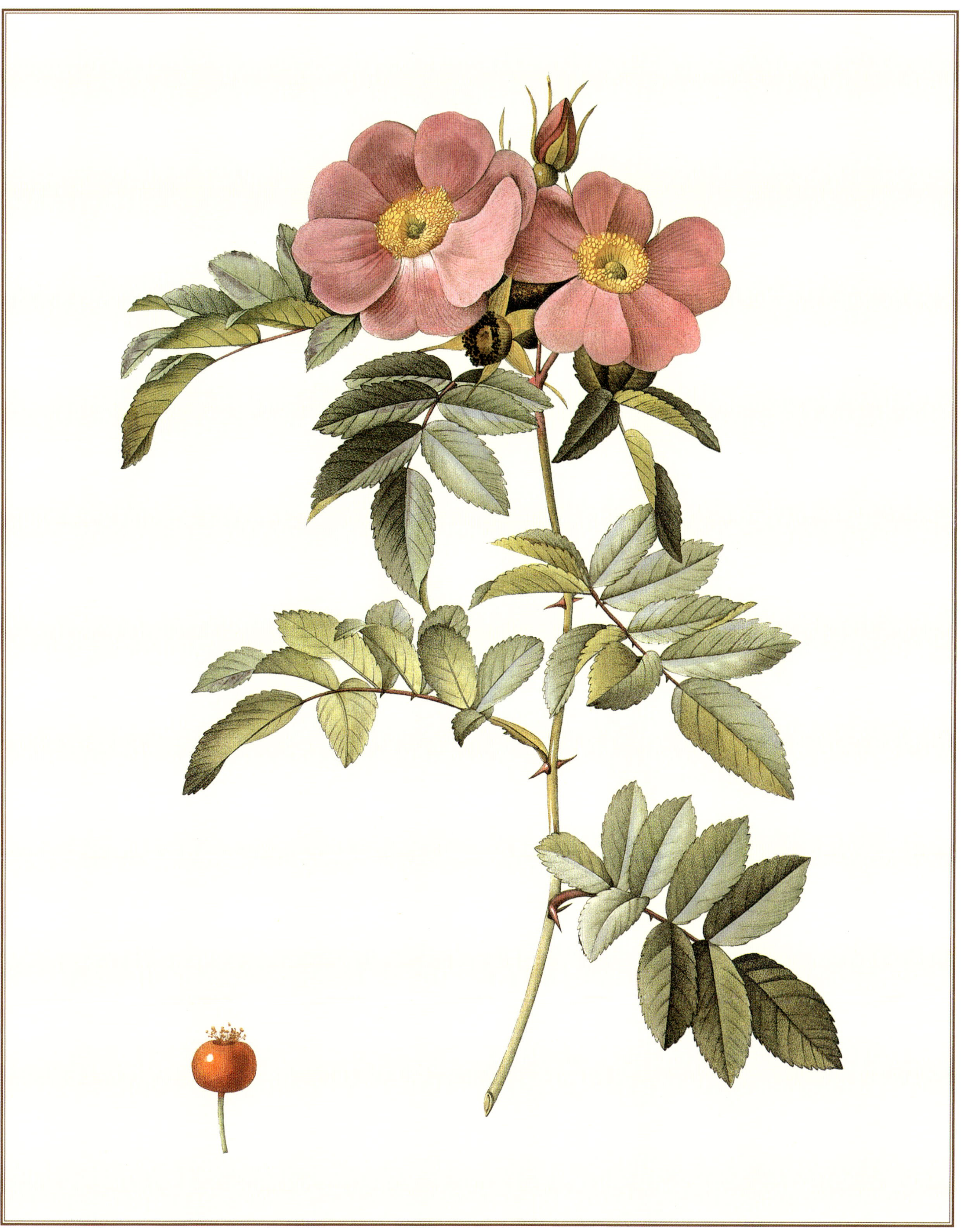

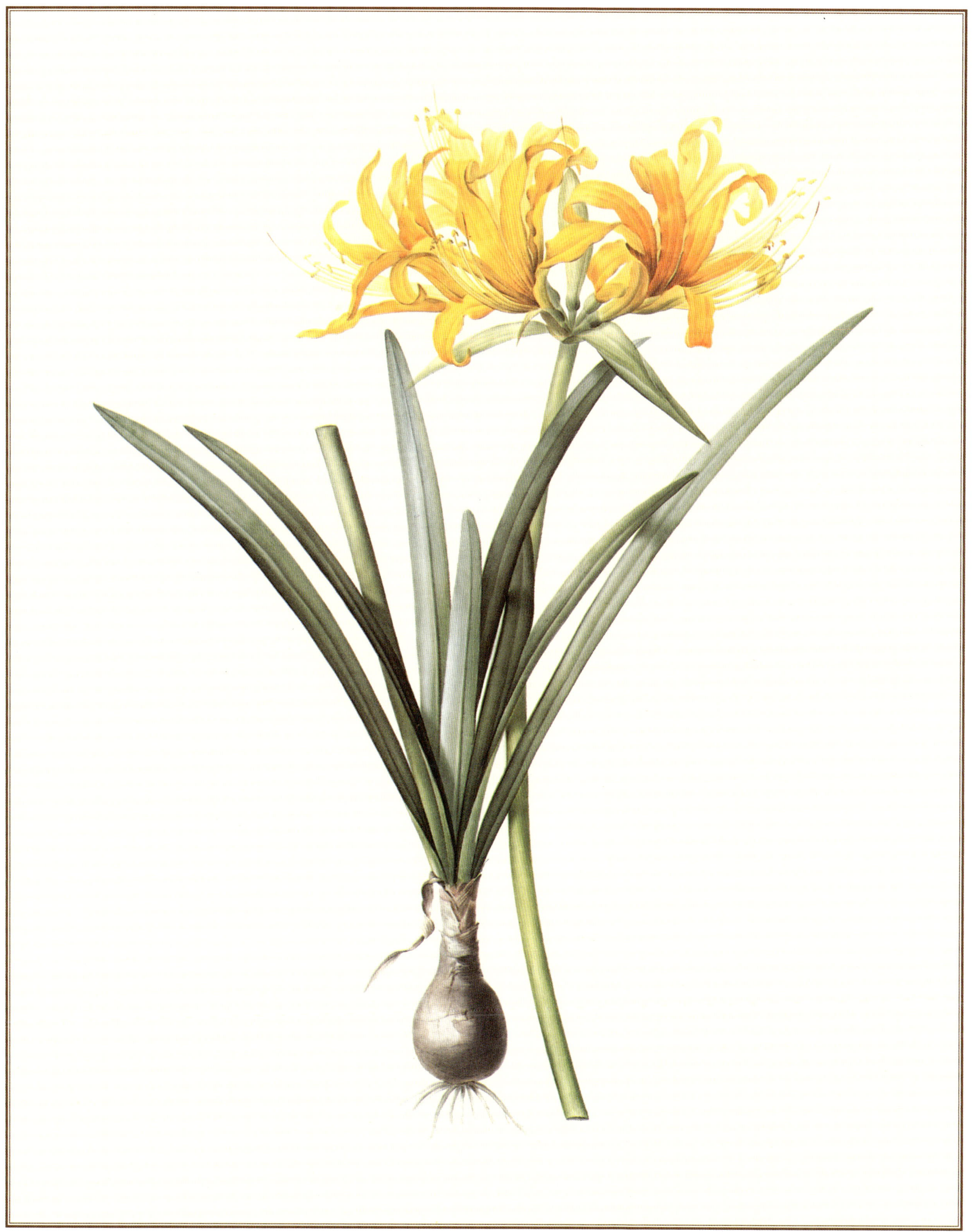

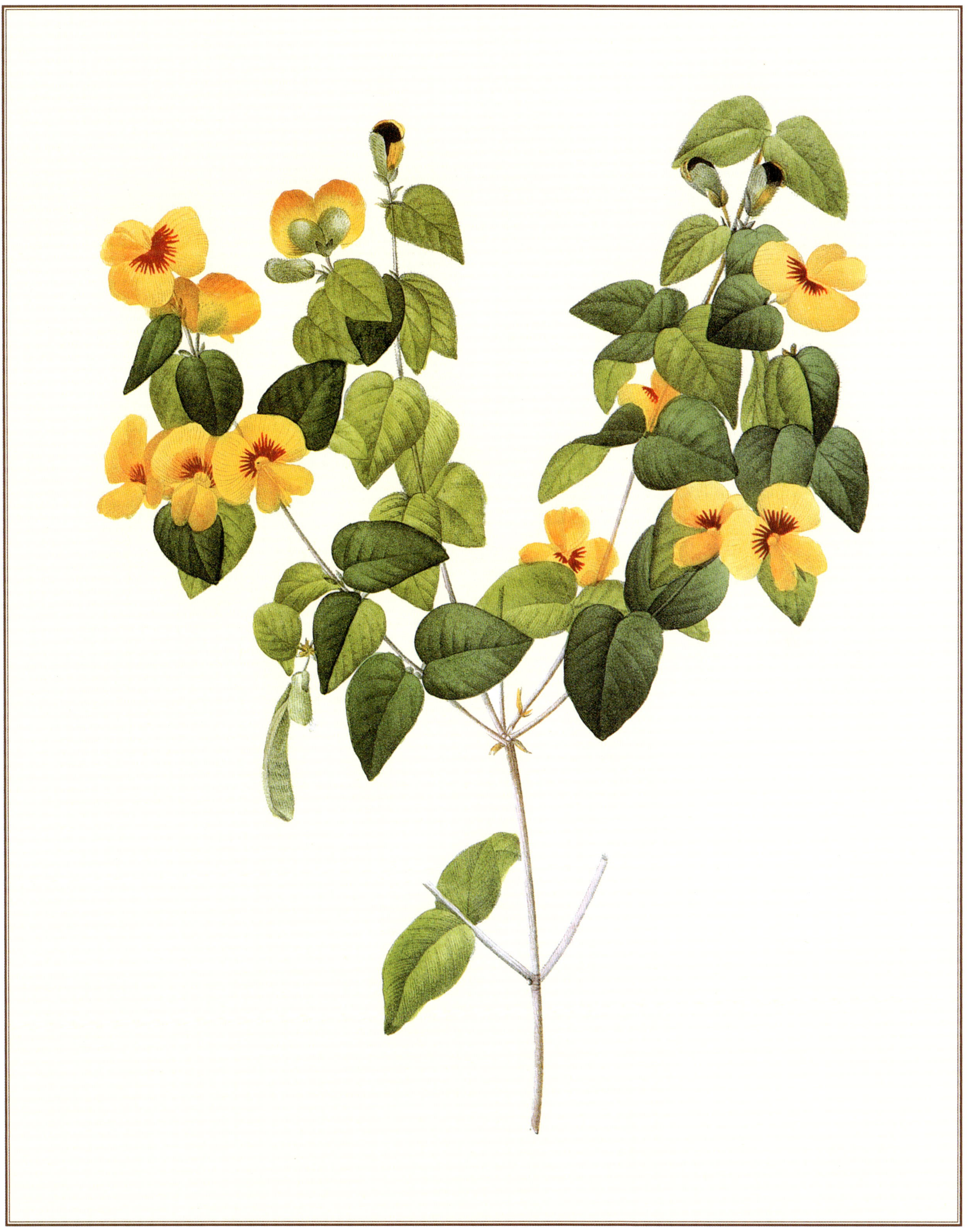

PLATE 44
HIPPEASTRUM
(*Hippeastrum puniceum*)

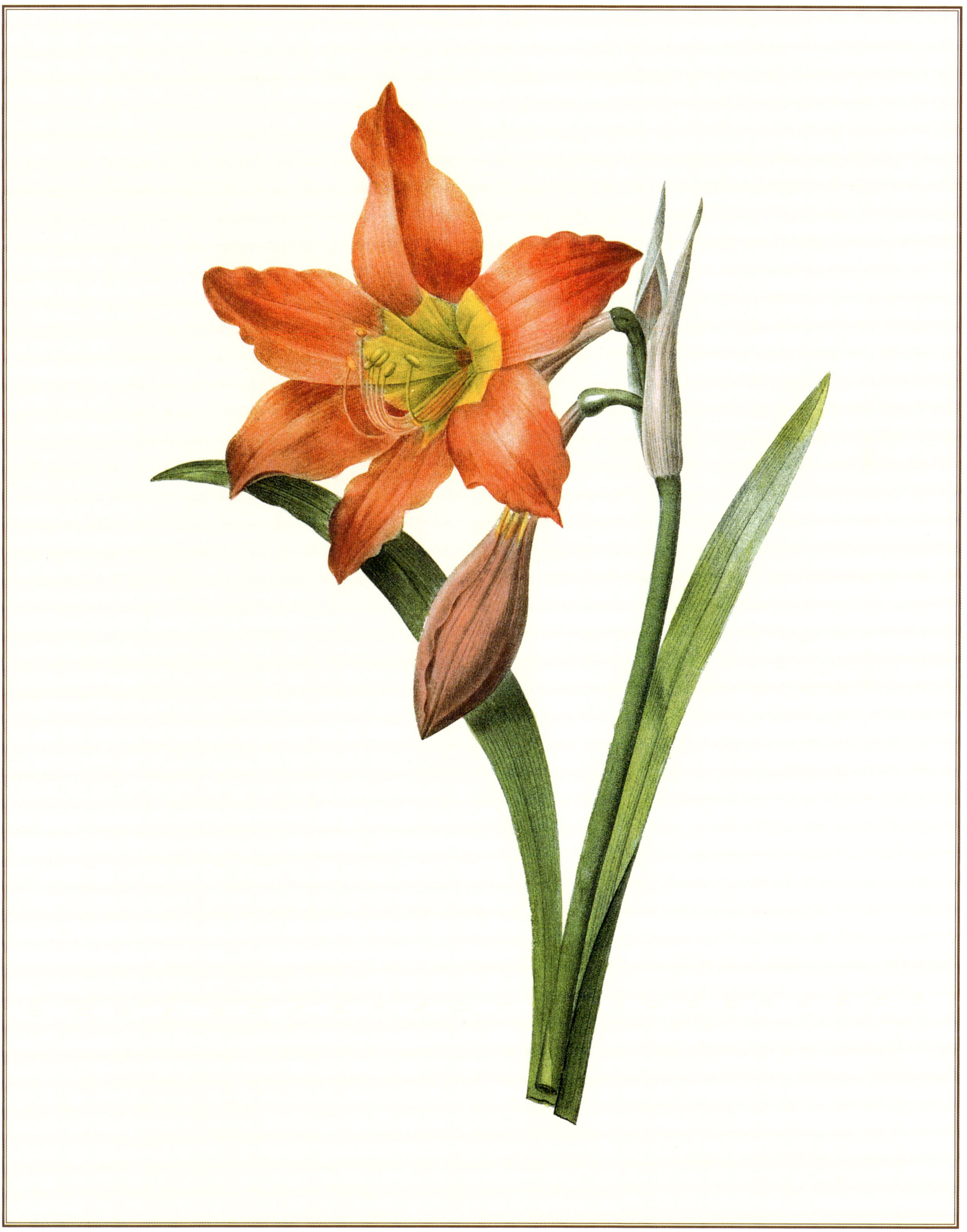

PLATE 45
HOSTA
(*Hosta venticosa*)

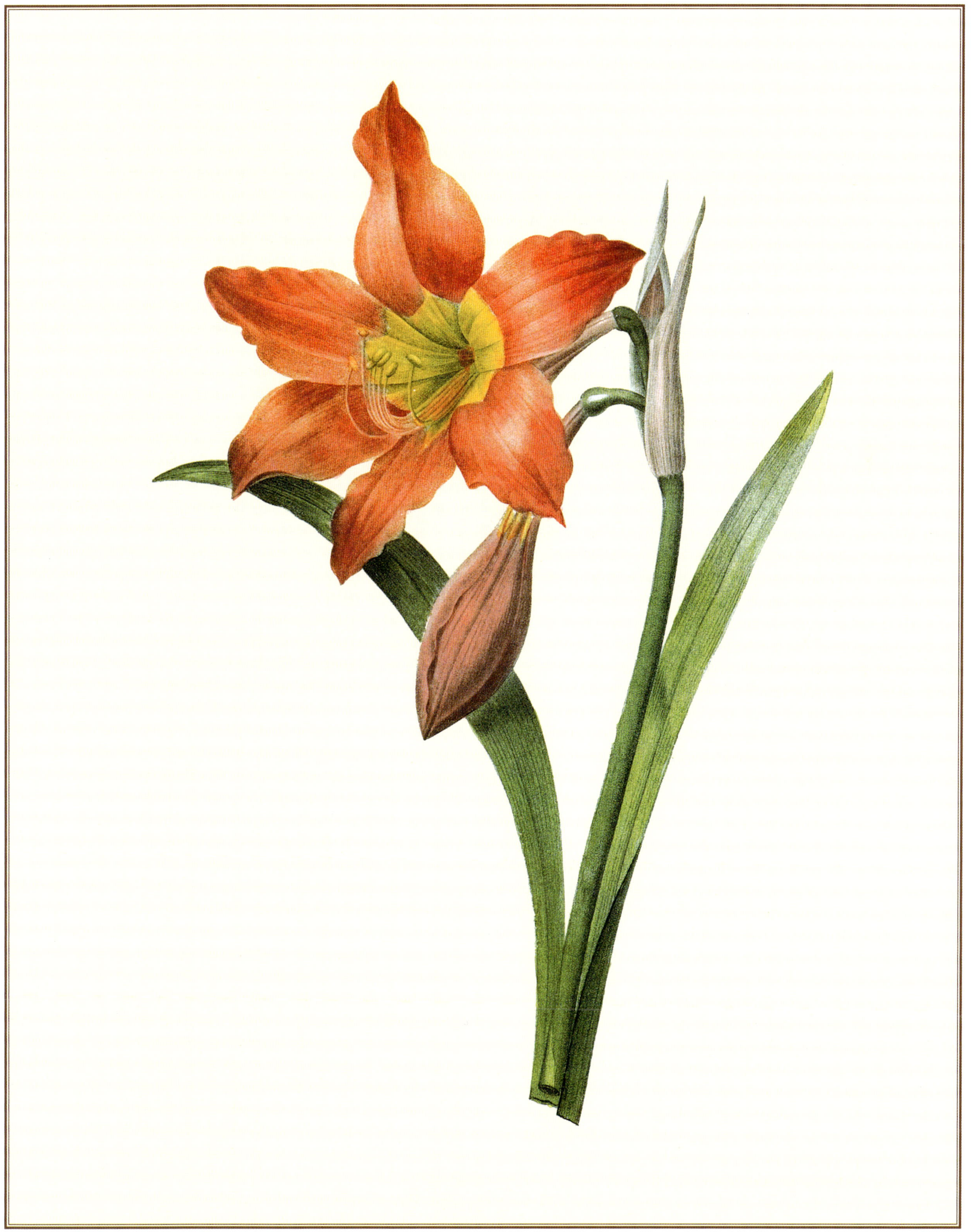

PLATE 45

HOSTA

(Hosta venticosa)

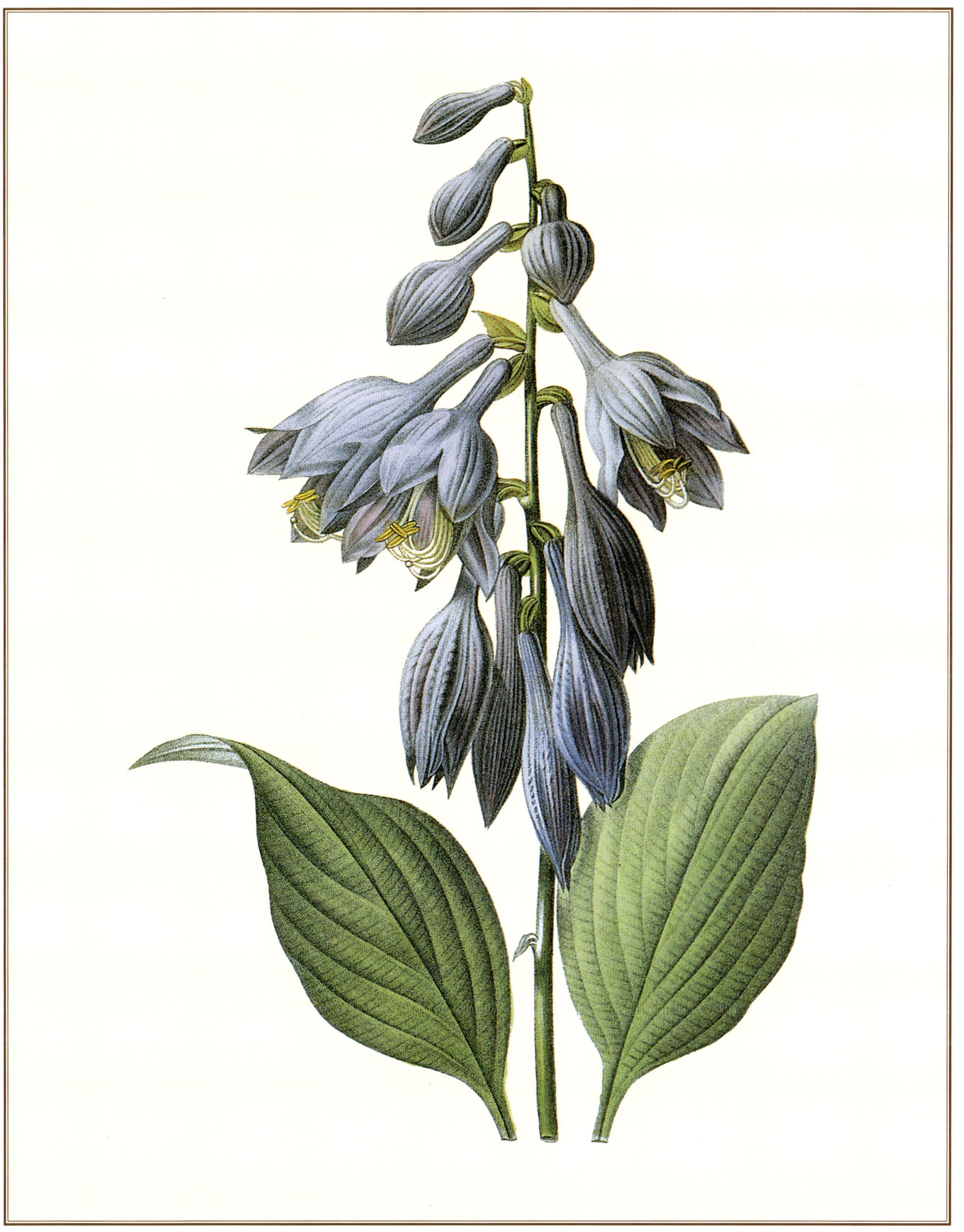

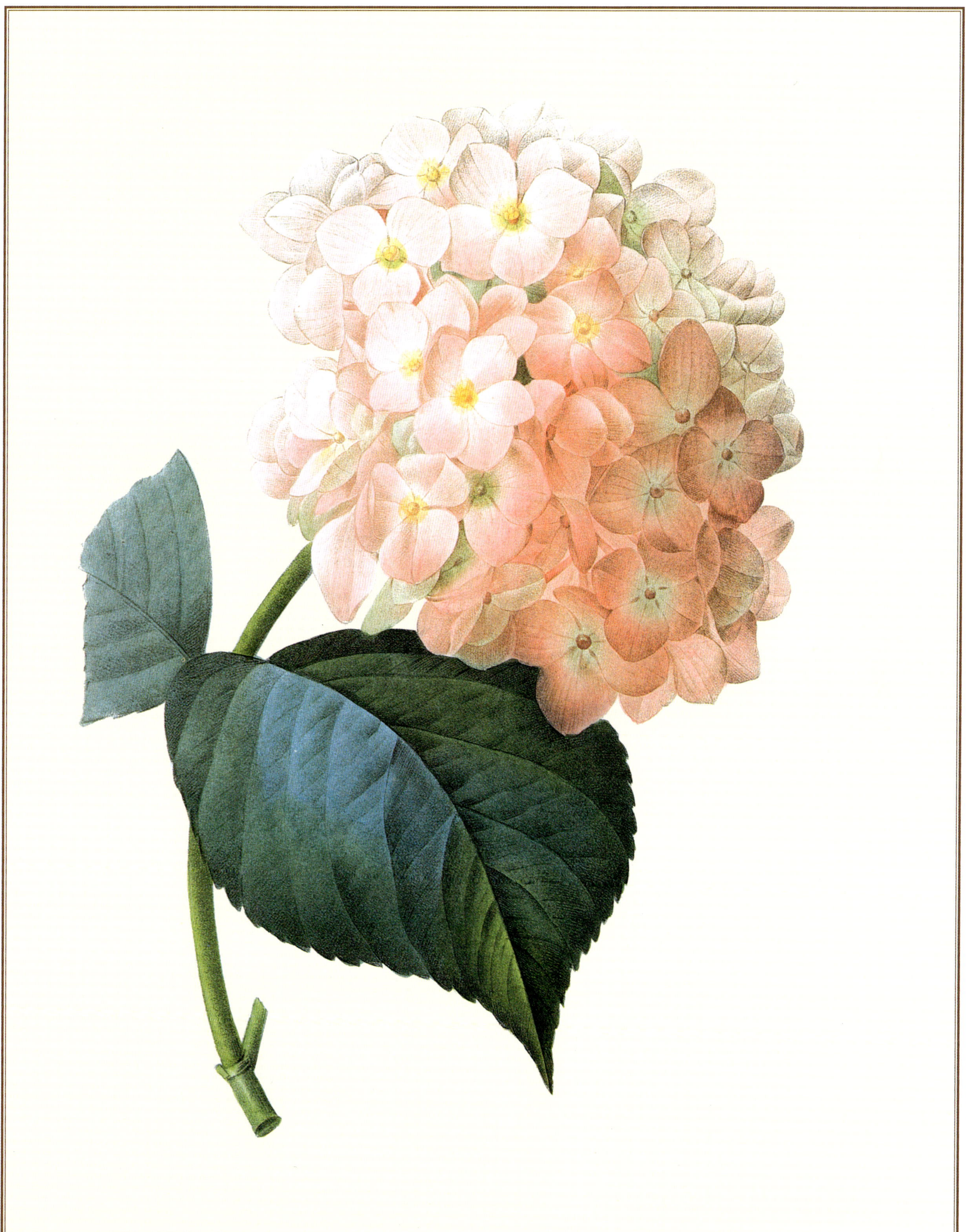

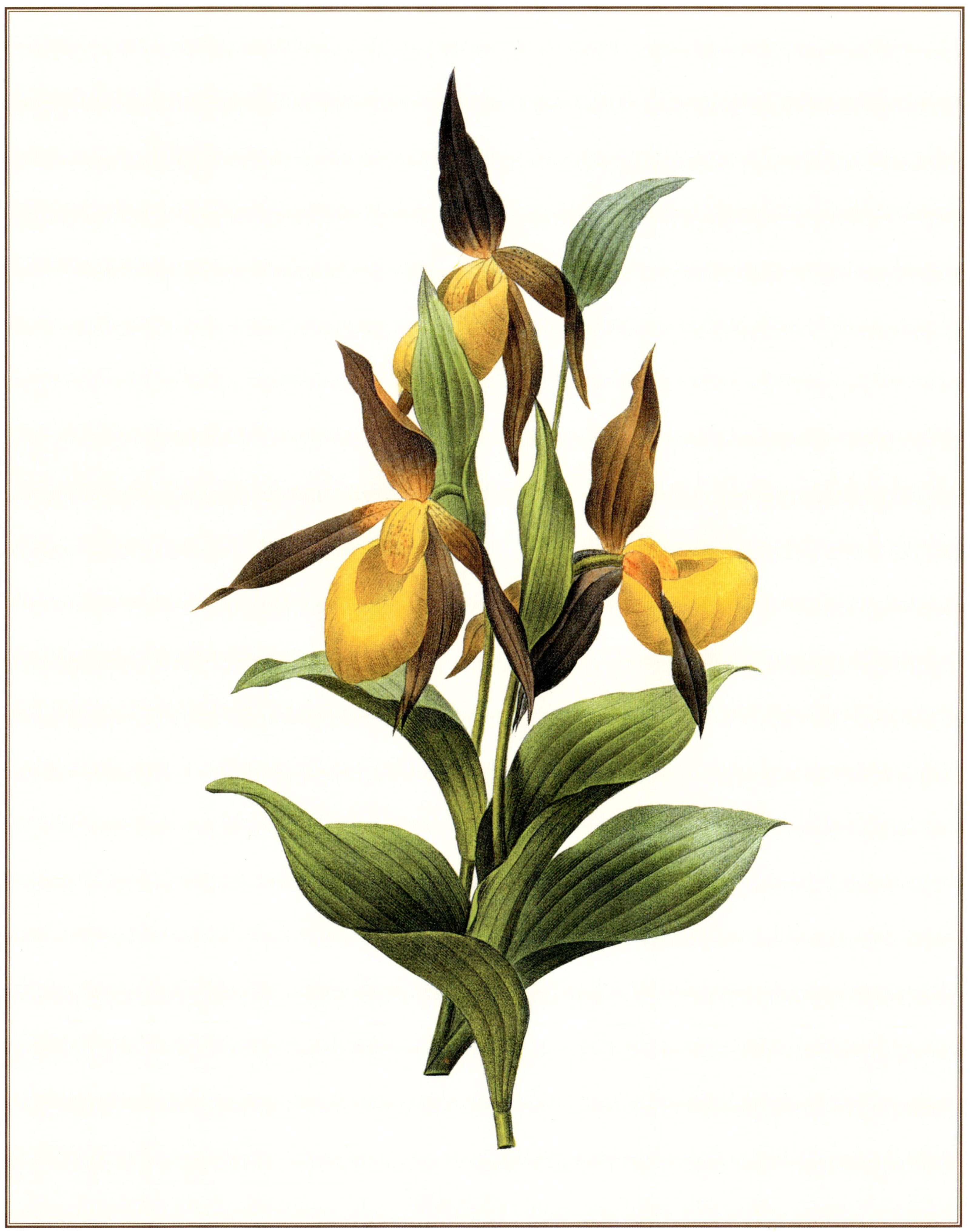

PLATE 48

LEADWORT

(Plumbago auriculata)

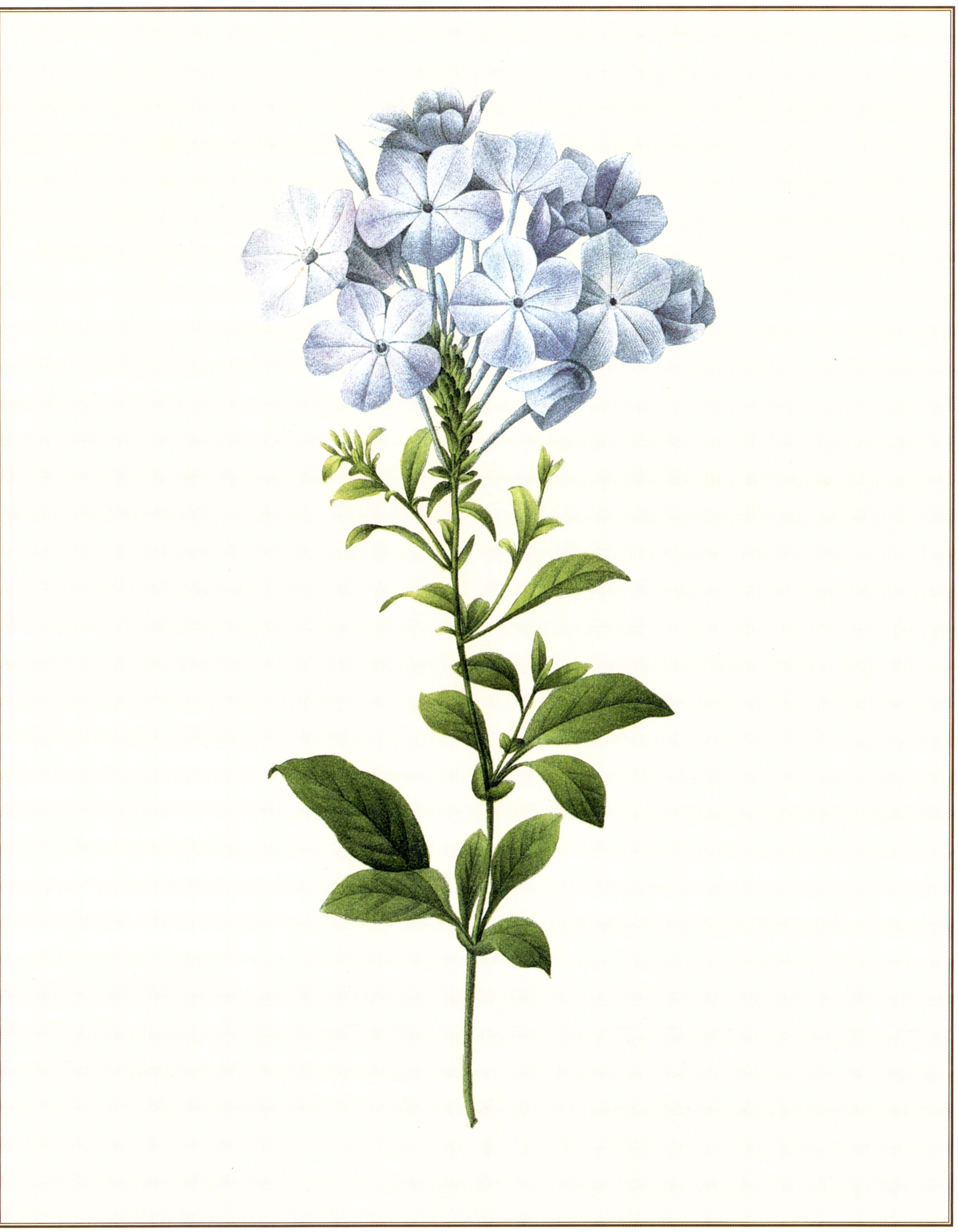

PLATE 49
L'HERITIER'S ROSE
(Rosa lheritieranea)

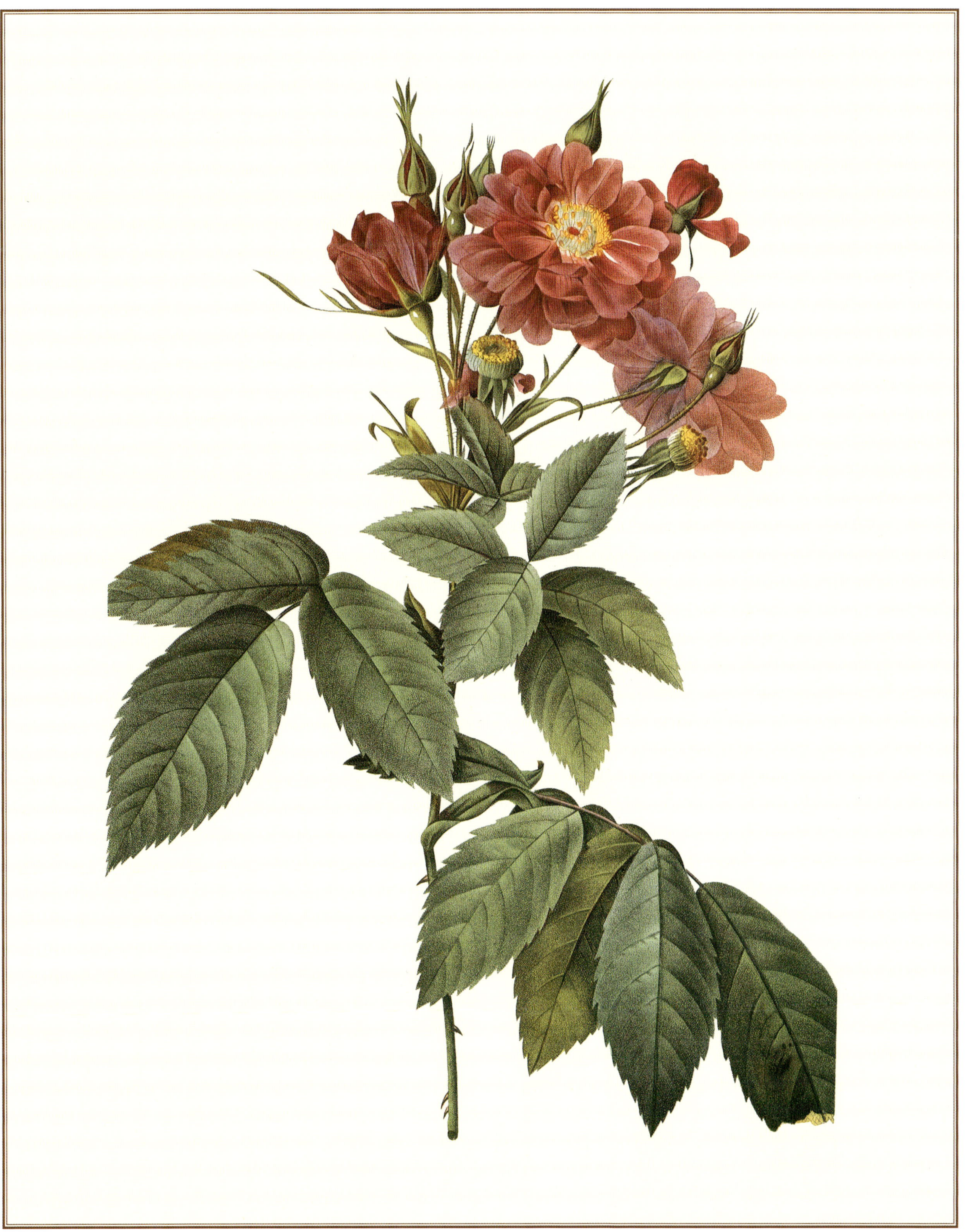

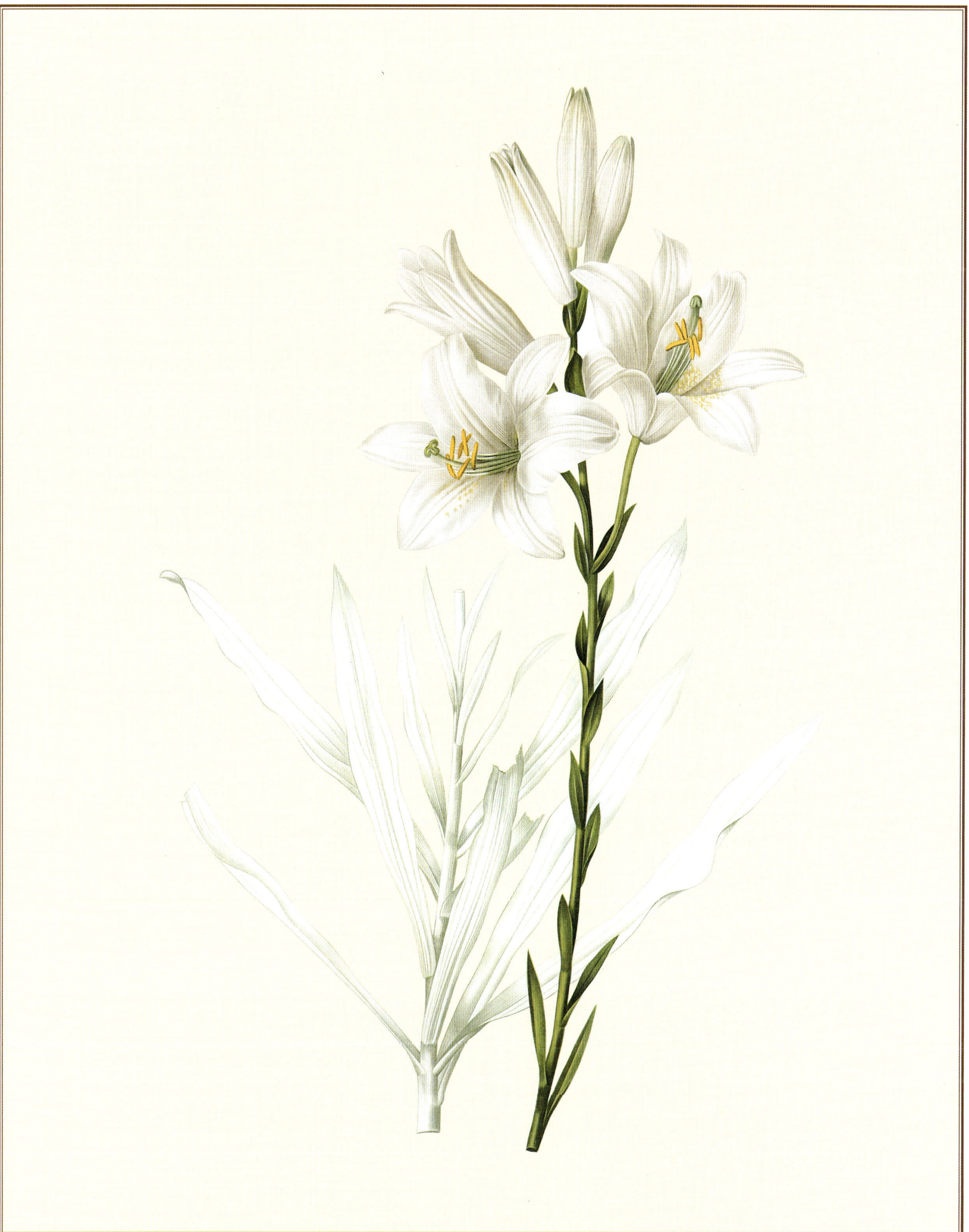

PLATE 51

MALMEDY ROSE

(*Rosa malmundariensis*)

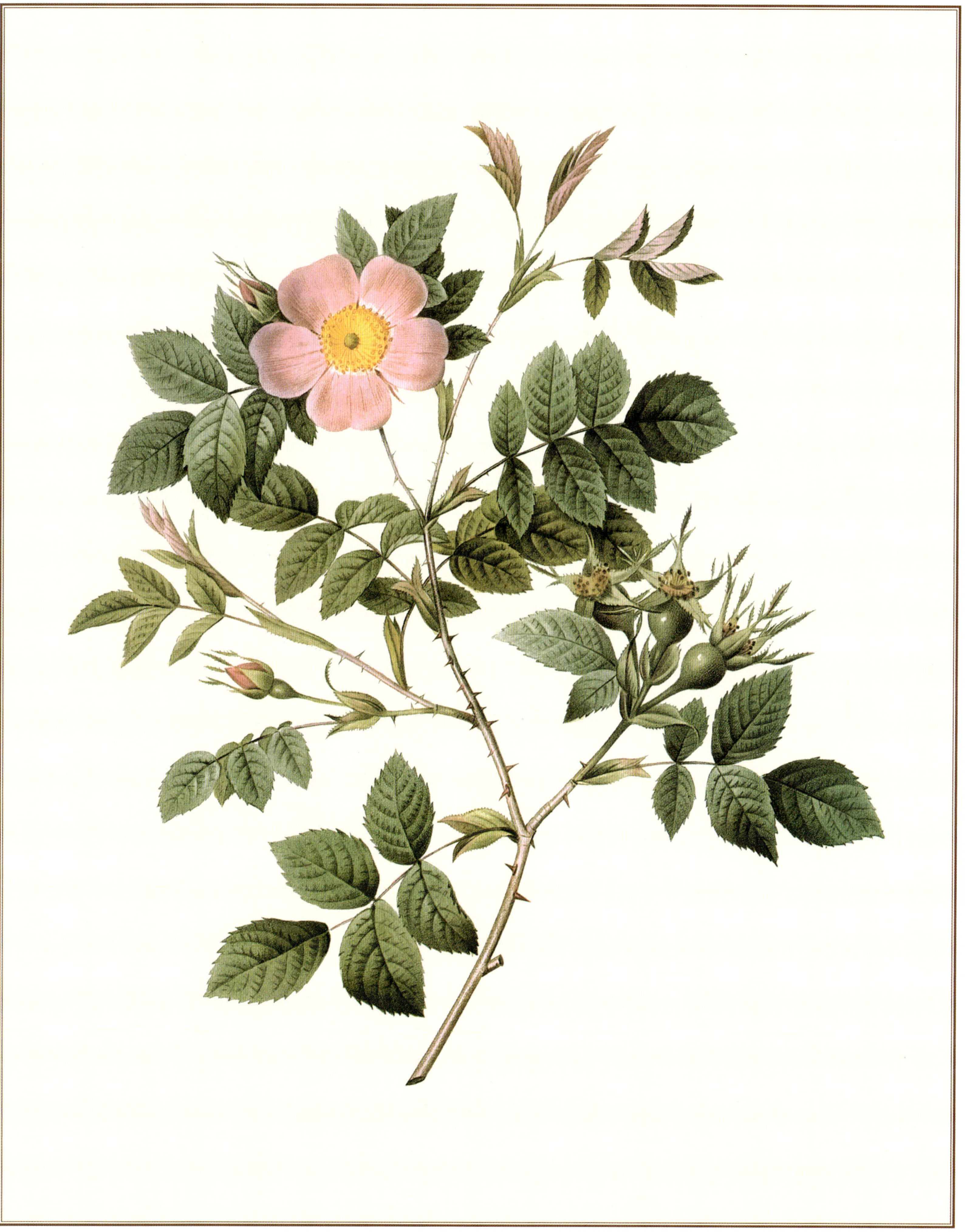

PLATE 52
MEADOW SAFFRON
(*Colchicum autumnale*)

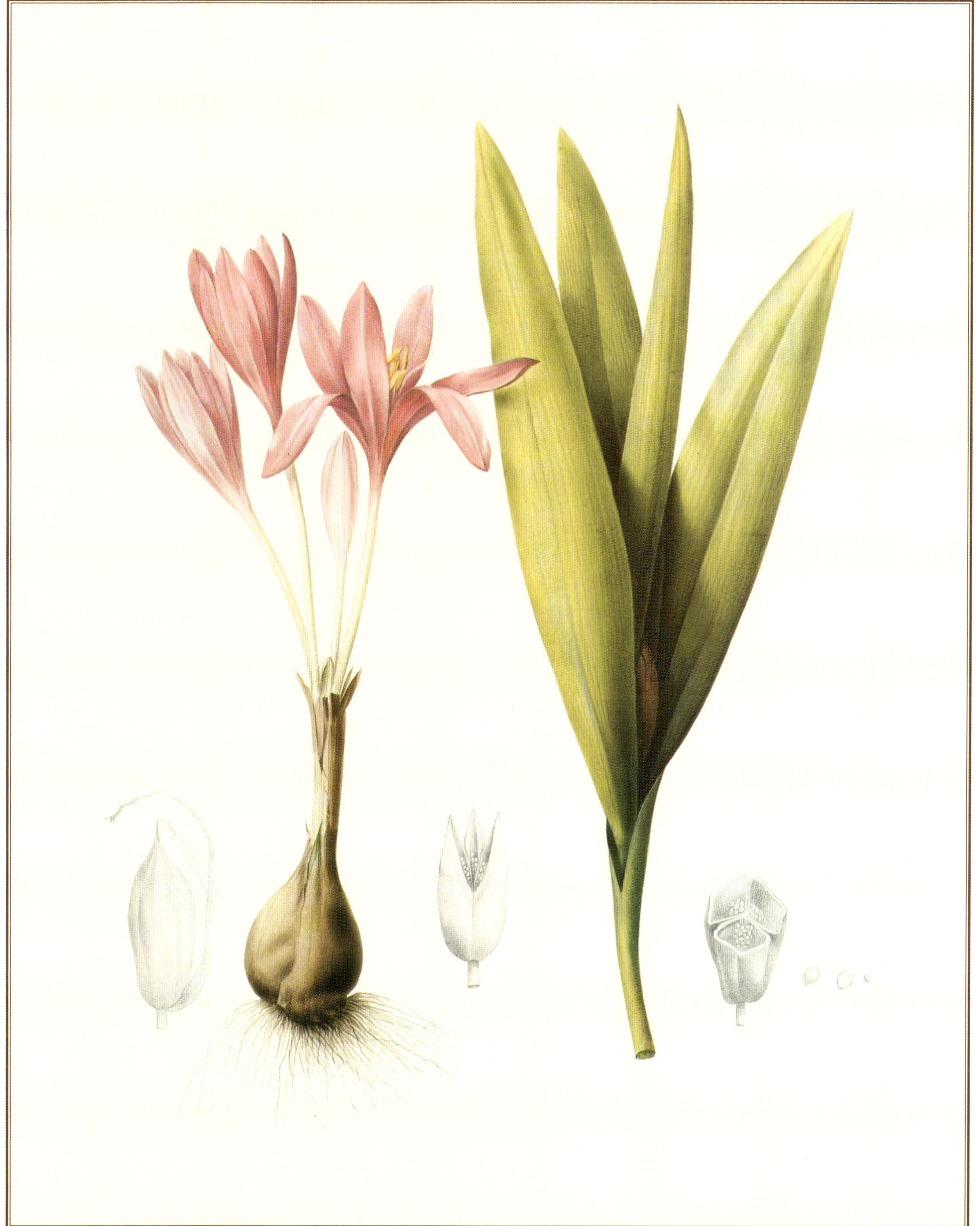

PLATE 53

MORNING GLORY

(Ipomoea quamoclit)

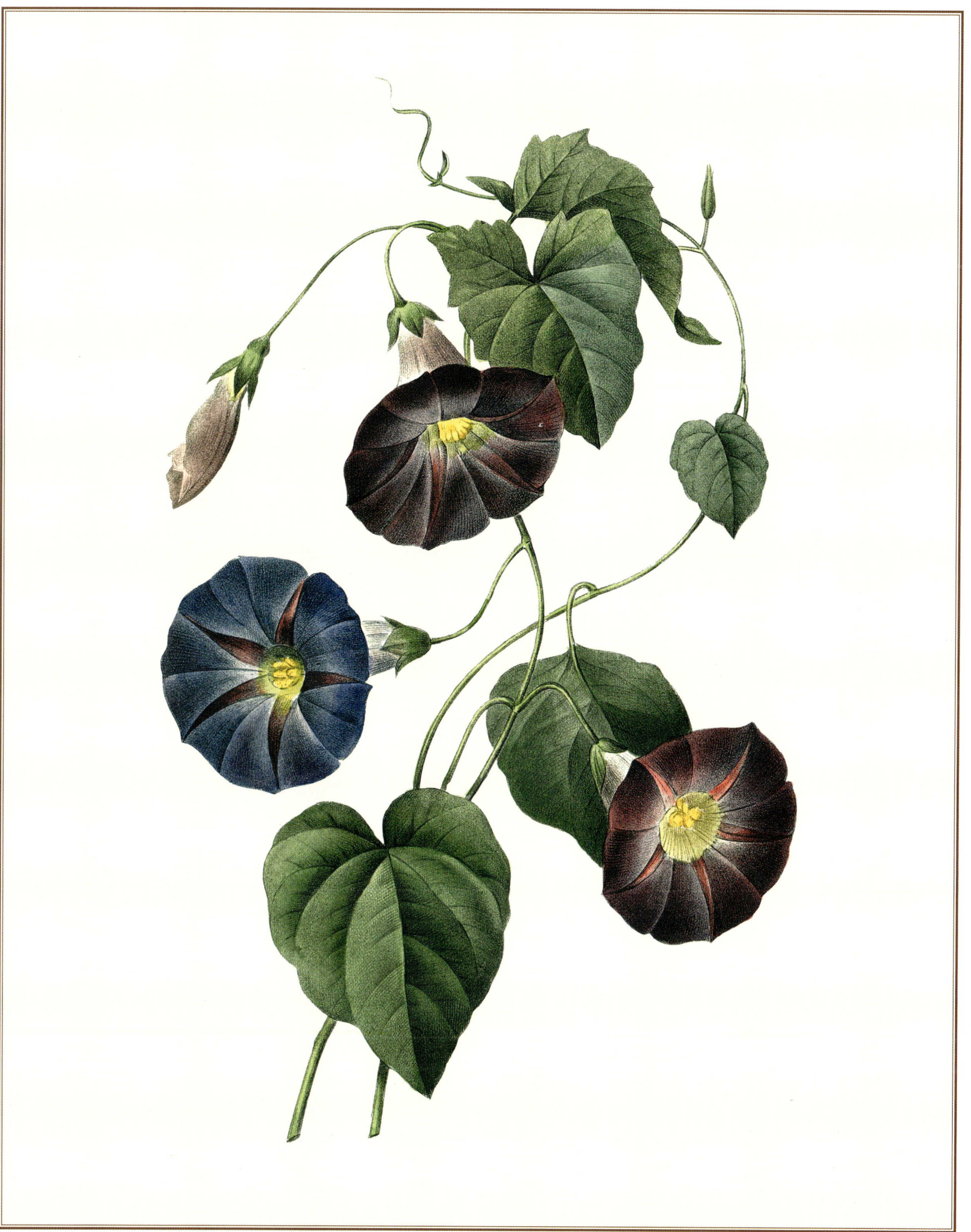

PLATE 54

OPIUM POPPY

(*Papaver somniferum*)

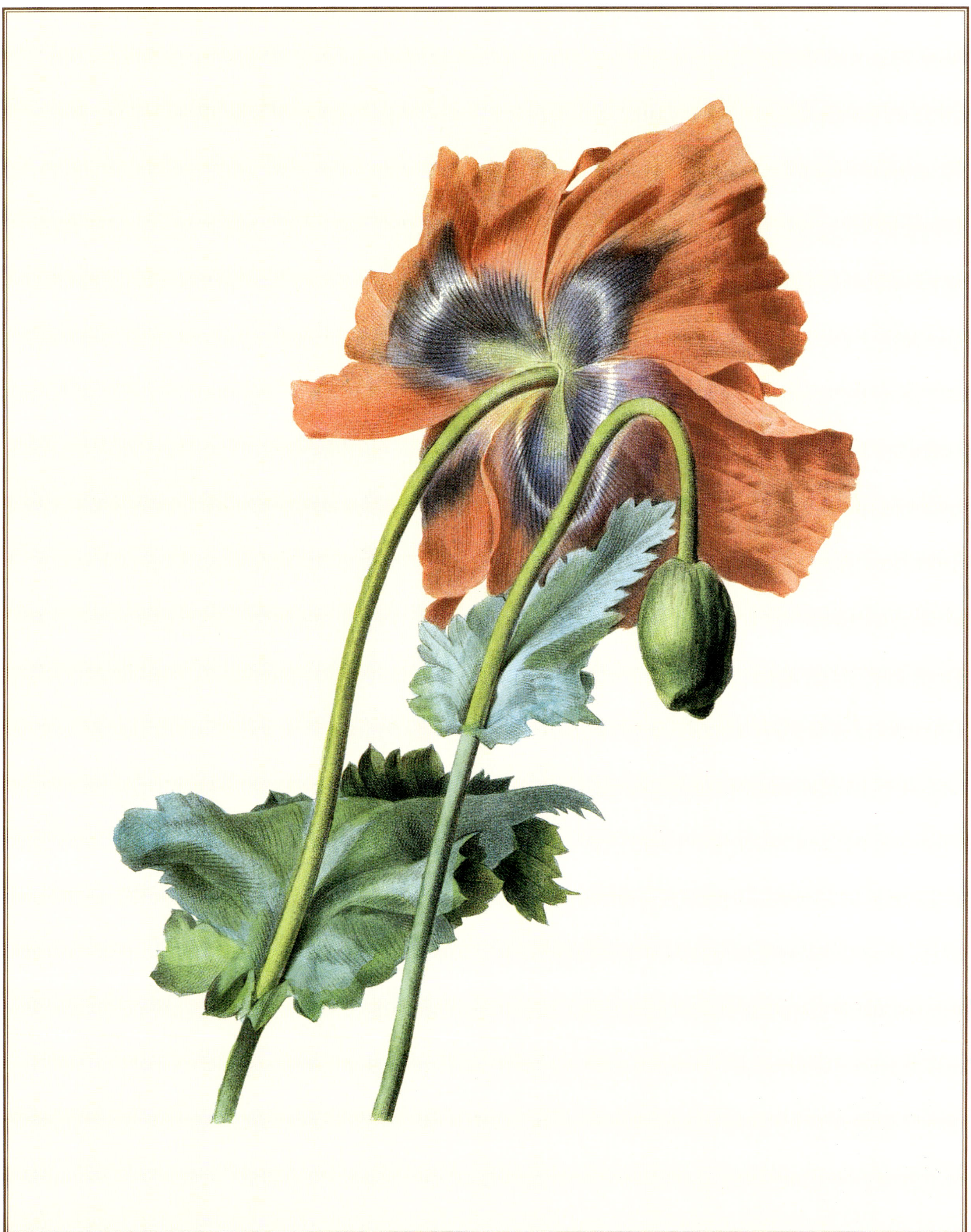

PLATE 55
ORANGE SHERBERT
(*Lychnis coranata*)

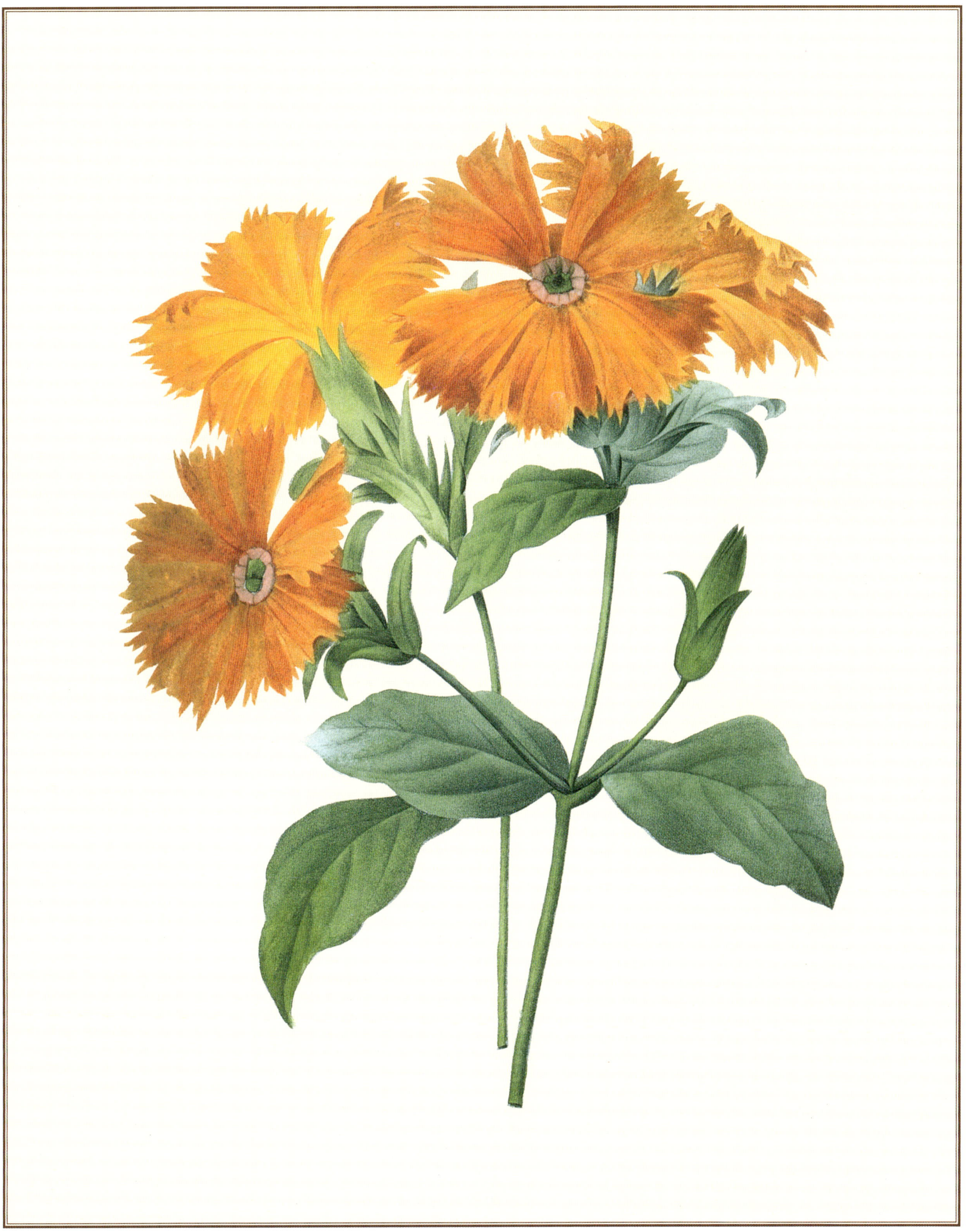

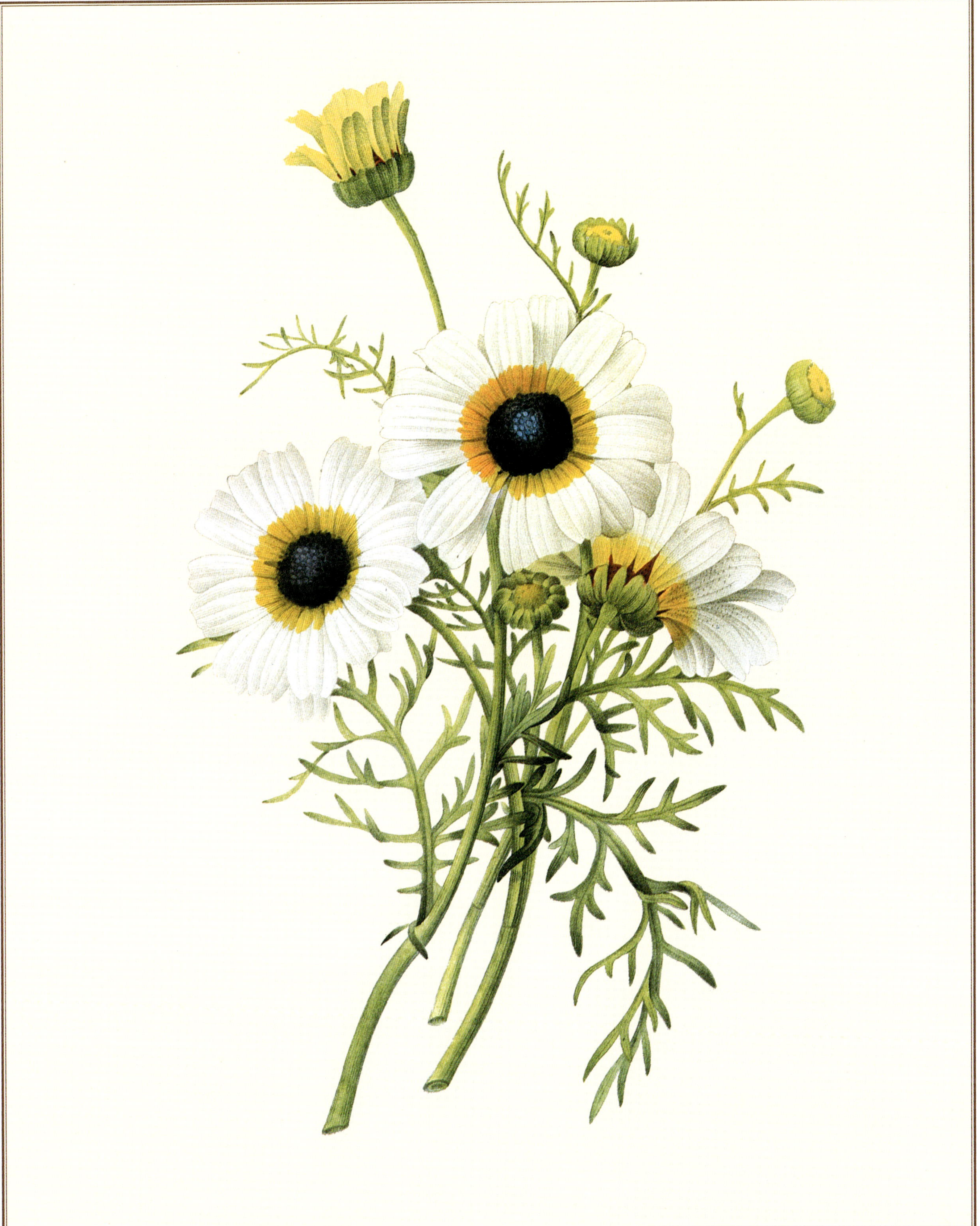

PLATE 57
PANSIES
(Viola tricolor)

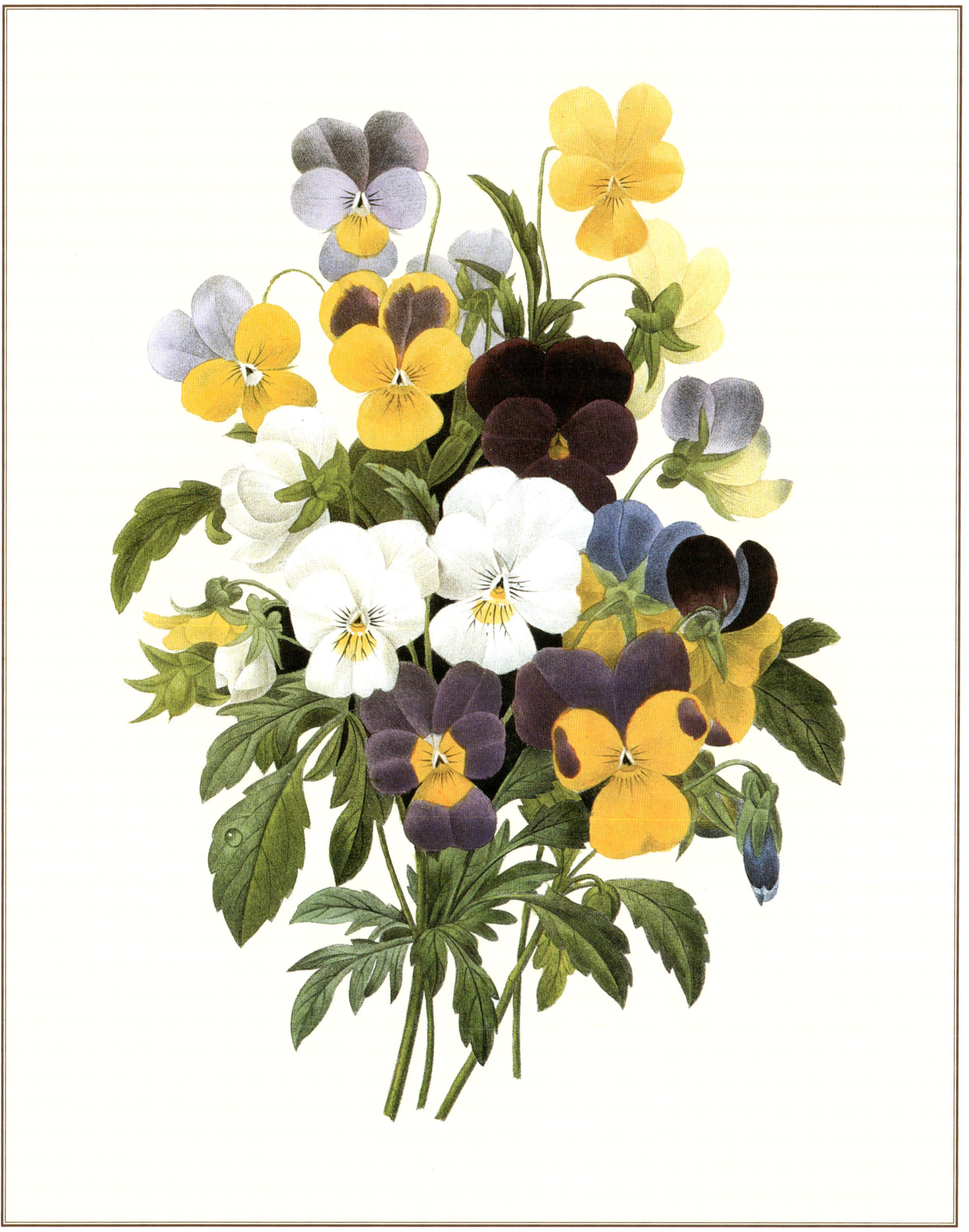

PLATE 58
PINEAPPLE
(*Bromelia ananas*)

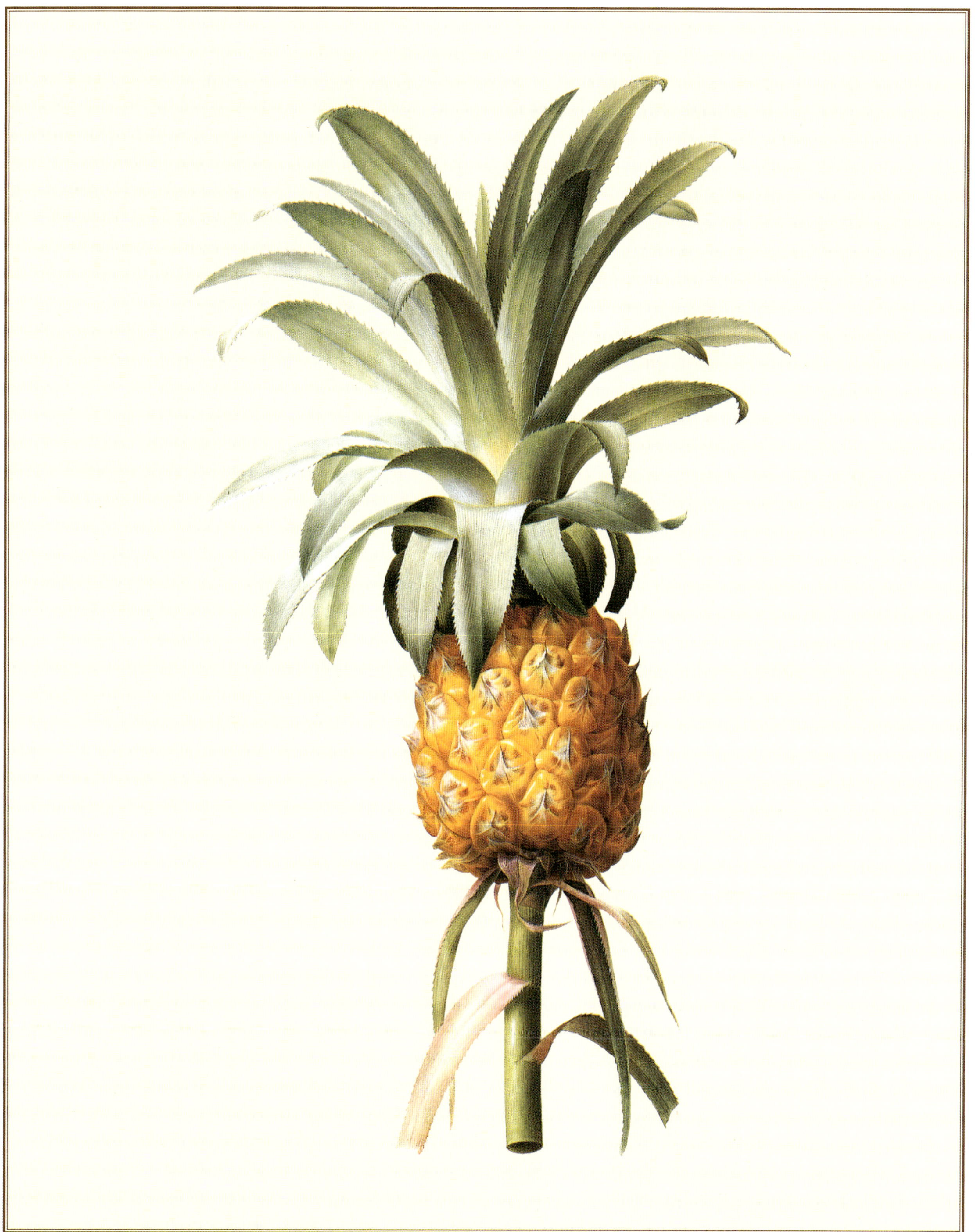

PLATE 59
PINK HEDGE ROSE
(*Rosa sepium rosea*)

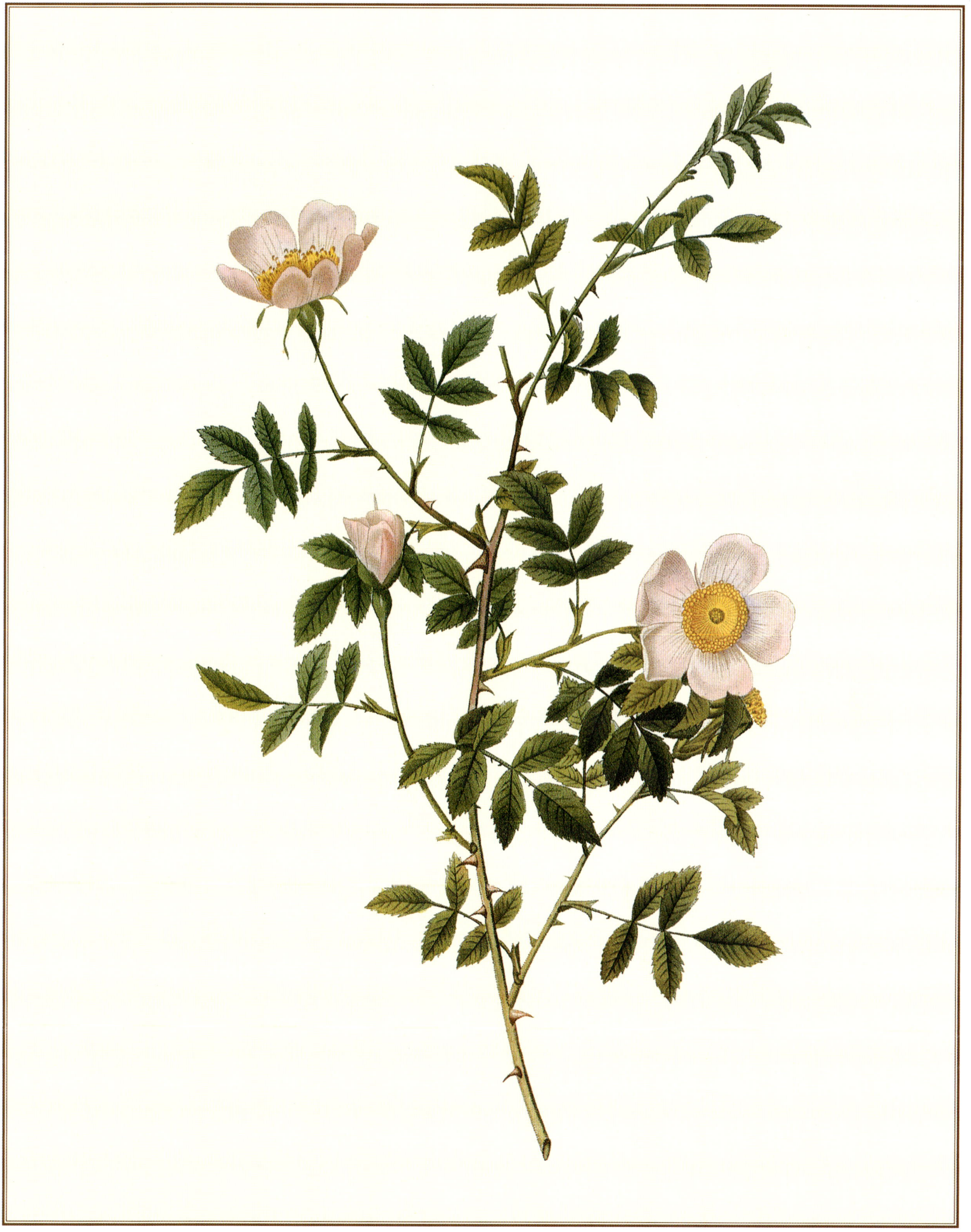

Plate 60

Portland Rose

(Rosa damascena coccinea)

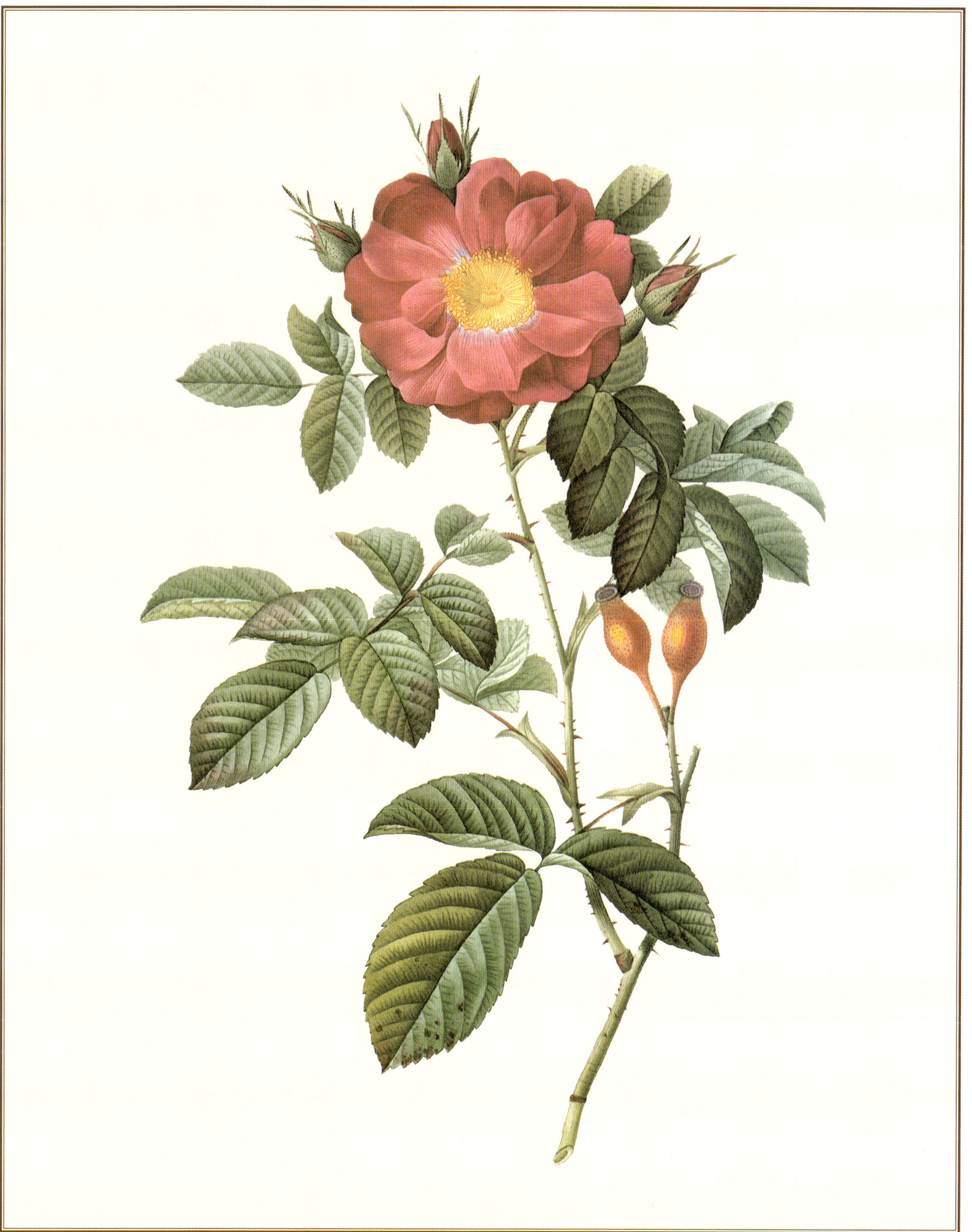

PLATE 61
PRIMROSE
(*Primula vulgaris*)

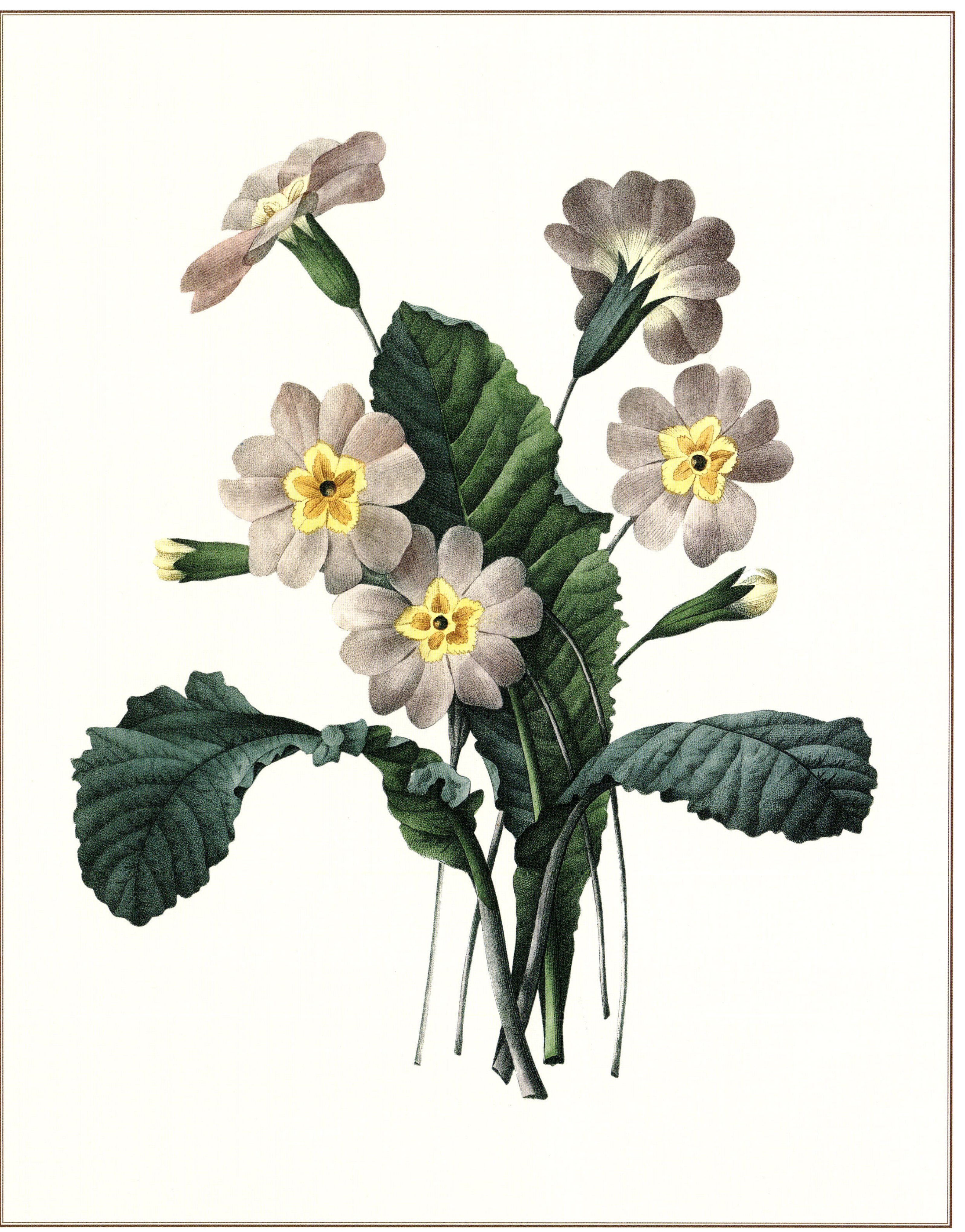

Plate 62
Red Giant
(*Amaryllis vittata*)

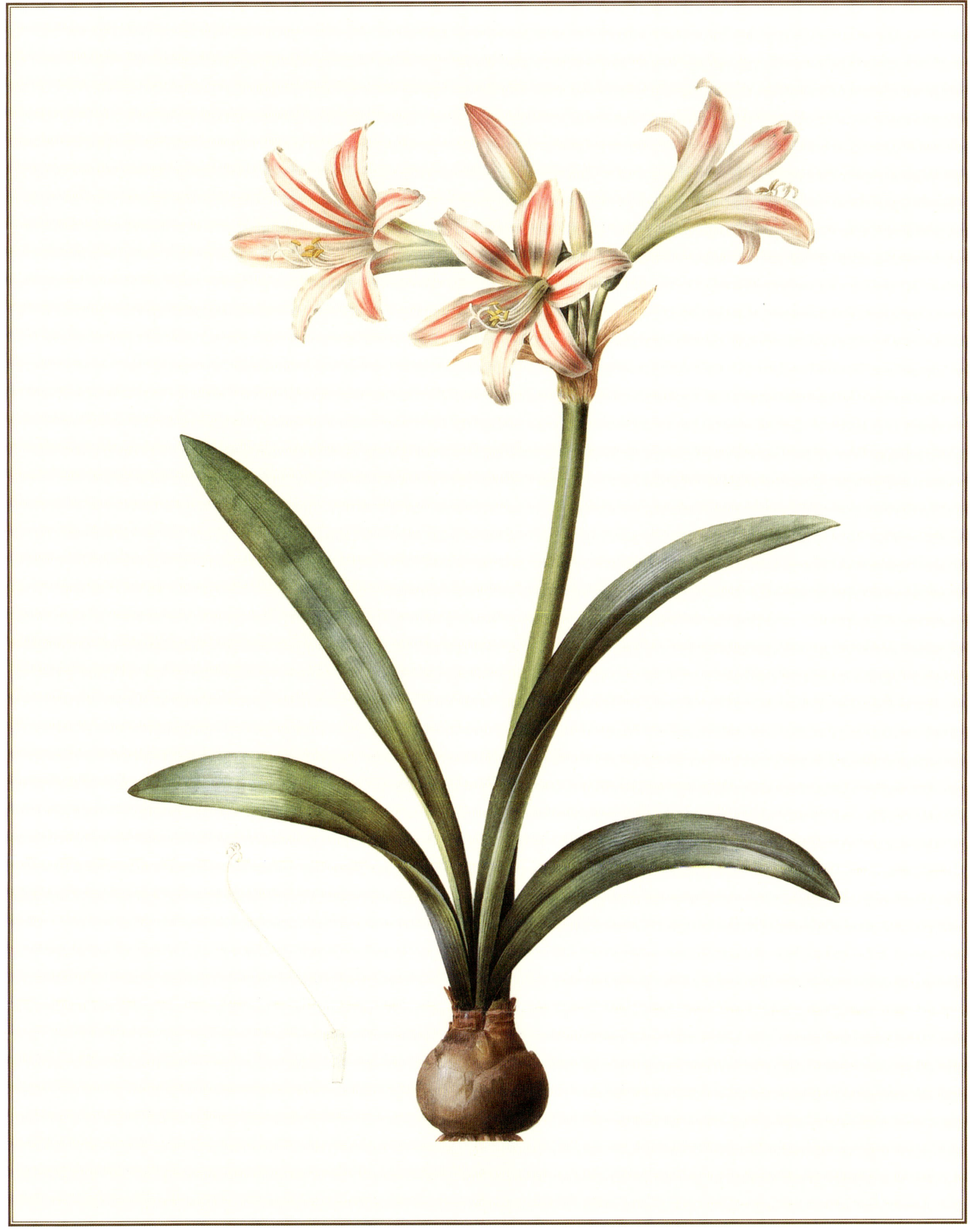

PLATE 63
RED-HOT POKER
(*Tritoma uvaria*)

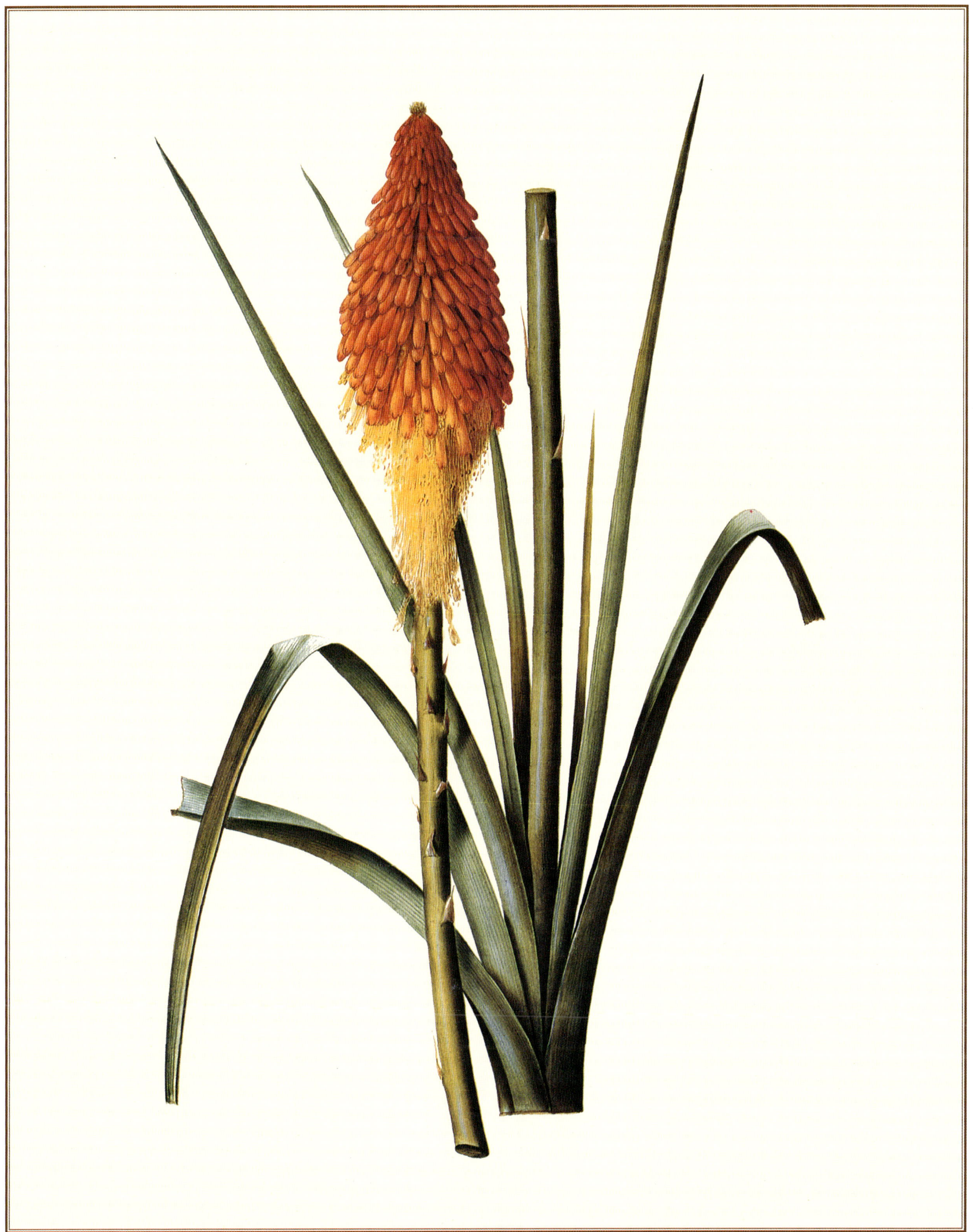

PLATE 64
RED-LEAVED ROSE
(*Rosa rubrifolia*)

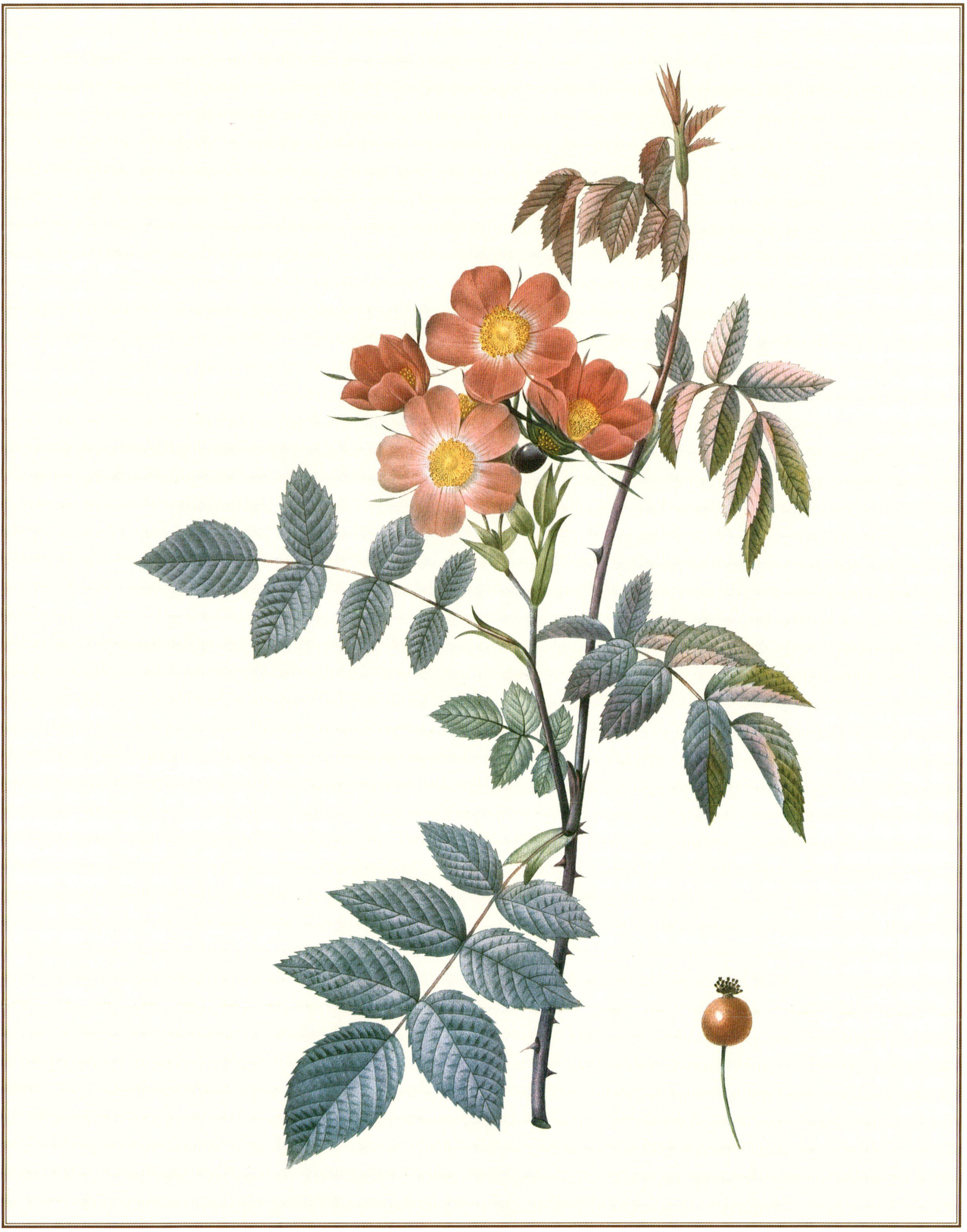

PLATE 65
ROSA MUNDI (STRIPED ROSE OF FRANCE)
(*Rosa gallica versicolor*)

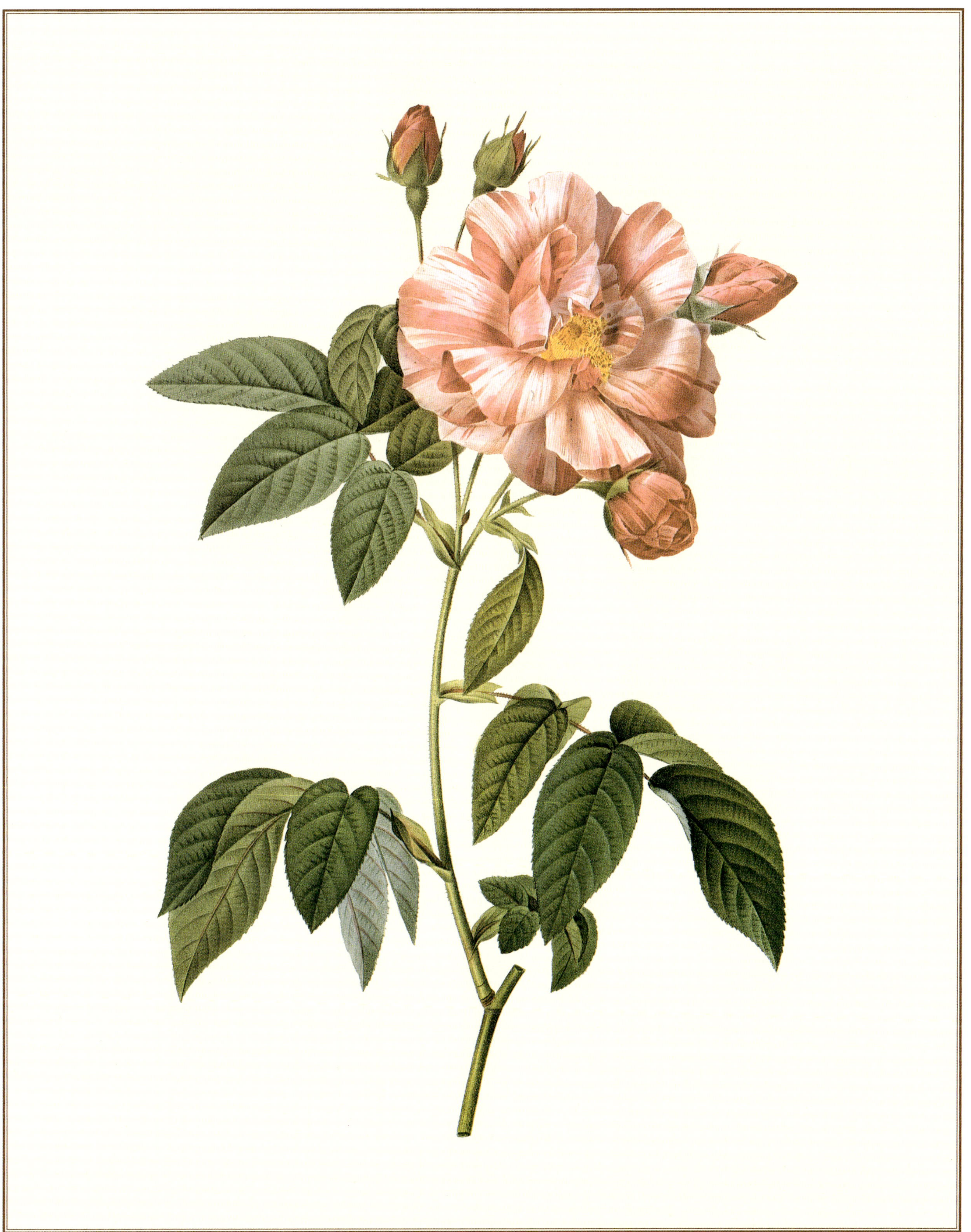

PLATE 66
ROSE OF LOVE
(*Rosa pumila*)

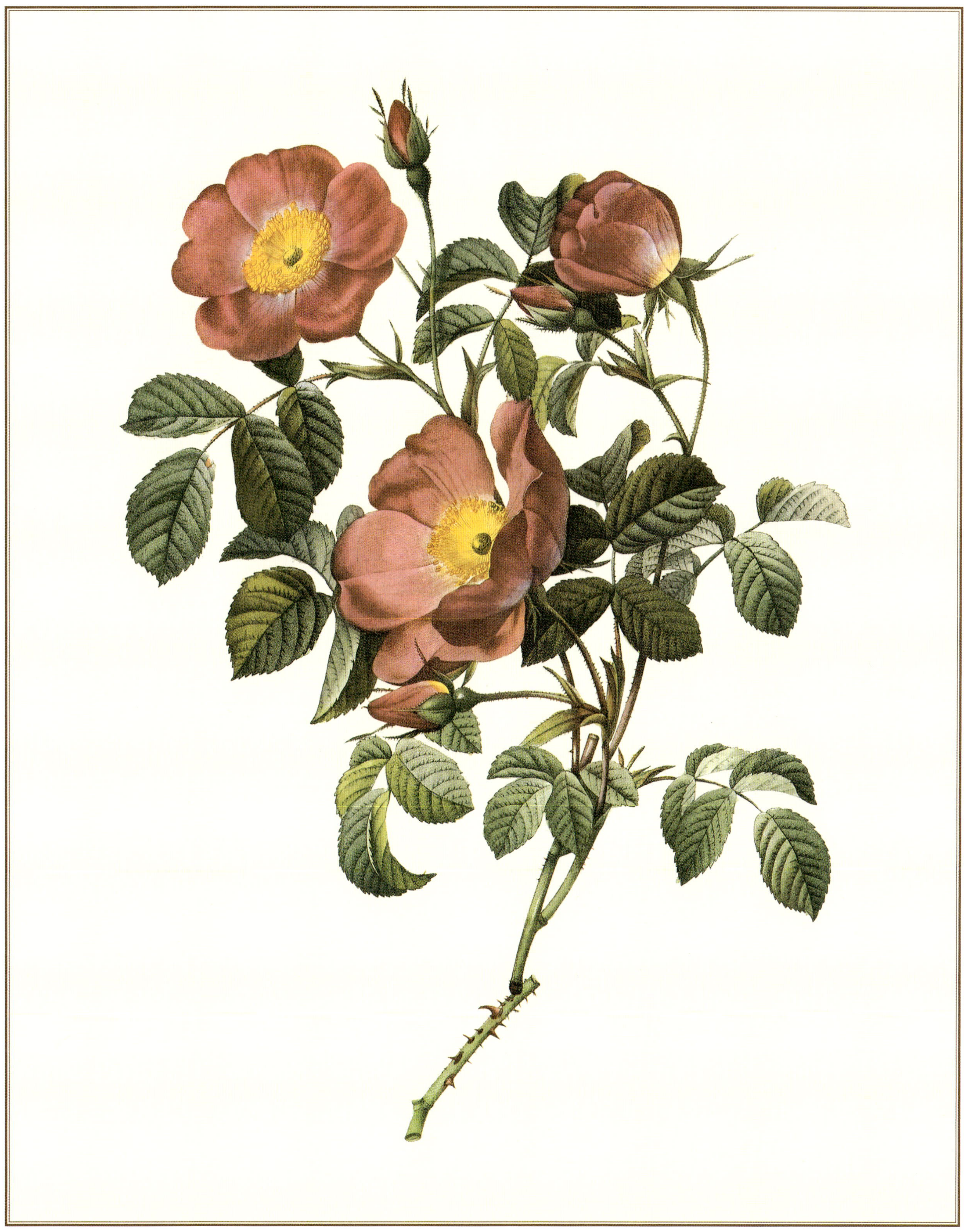

PLATE 67

ROSENBERG'S ROSE

(*Rosa rosenbergiana*)

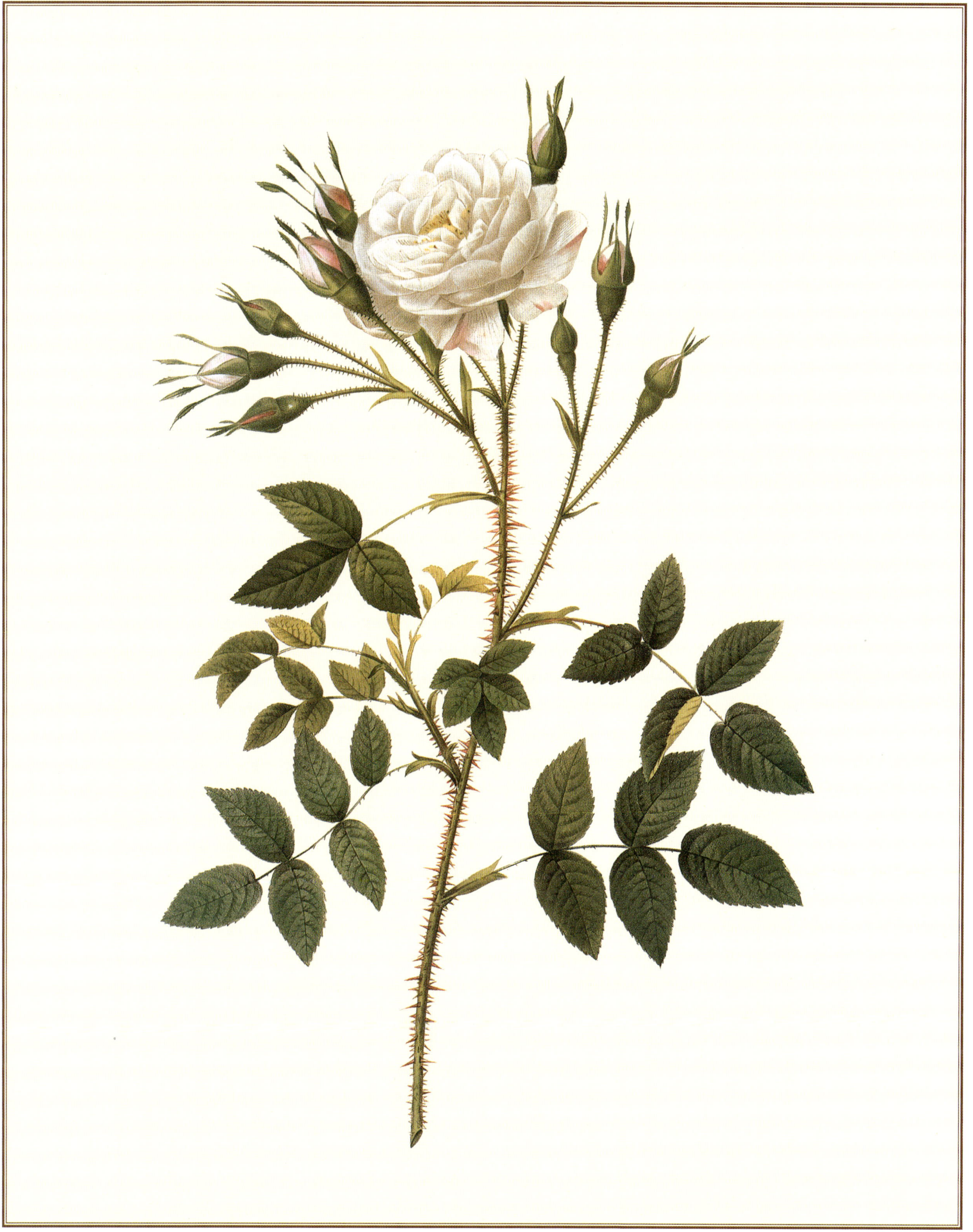

PLATE 68
SEMI-DOUBLE MUSK ROSE
(*Rosa moschata flore semi-pleno*)

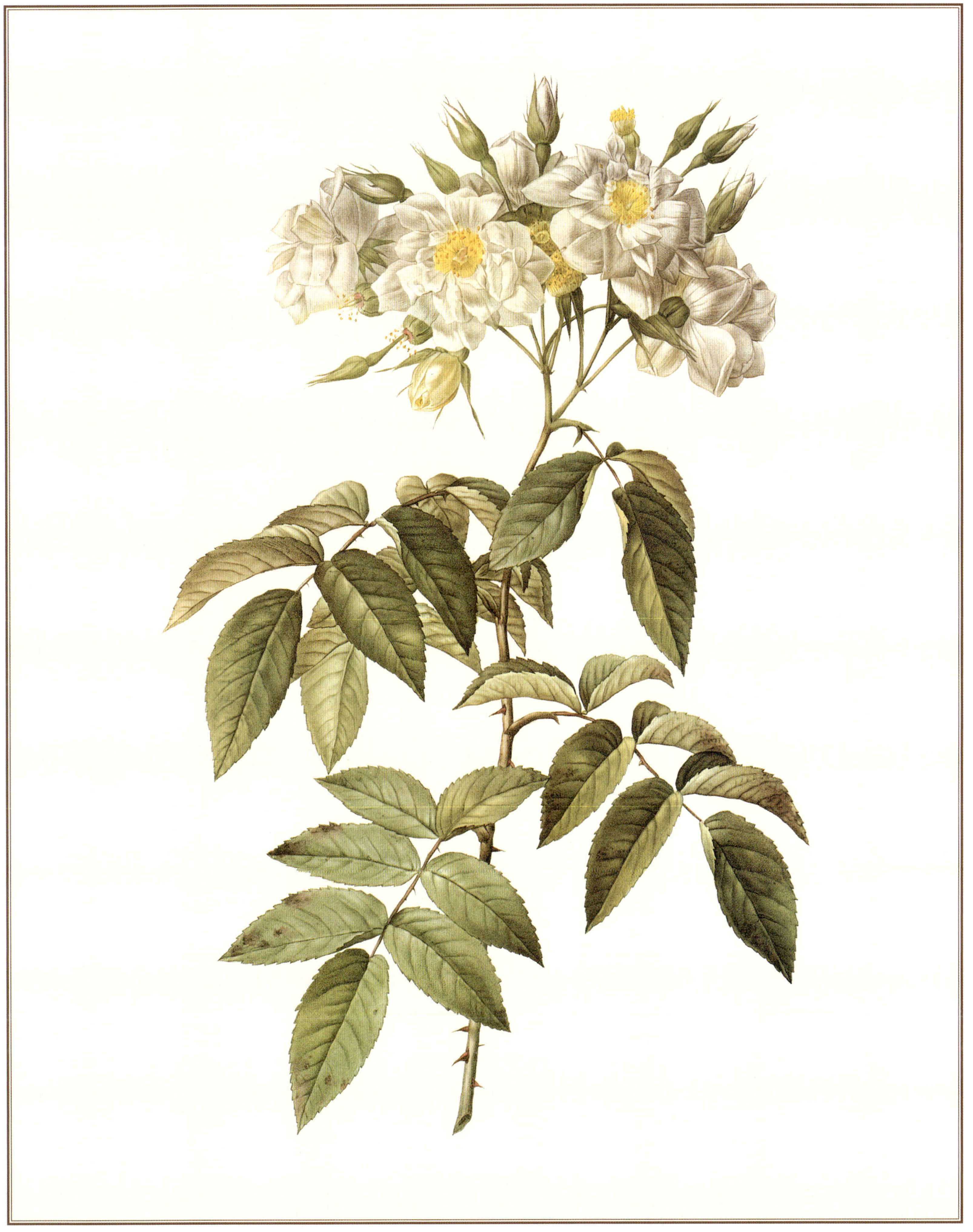

PLATE 69
SEMI-DOUBLE SWEETBRIAR
(*Rosa rubiginosa flore semi-pleno*)

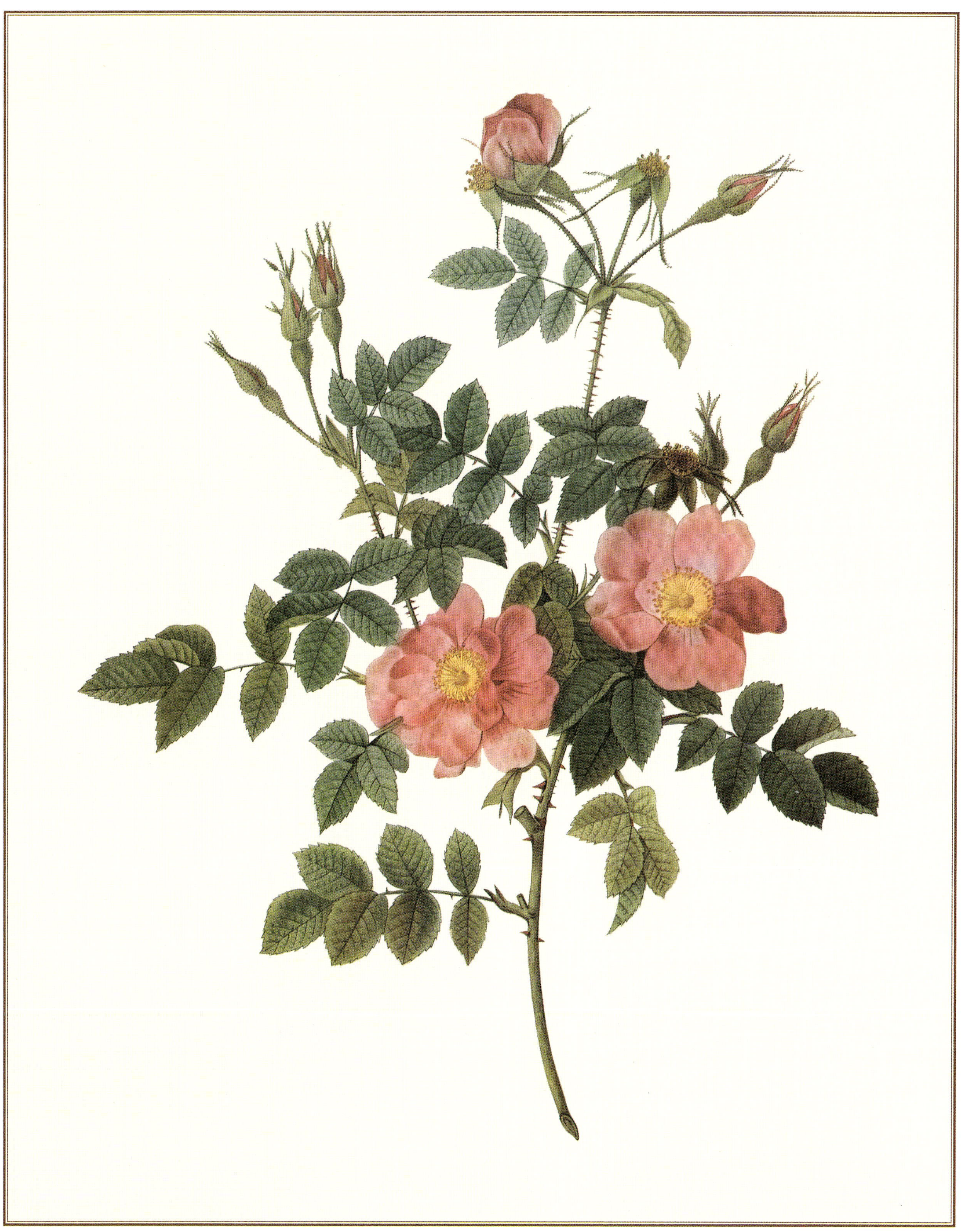

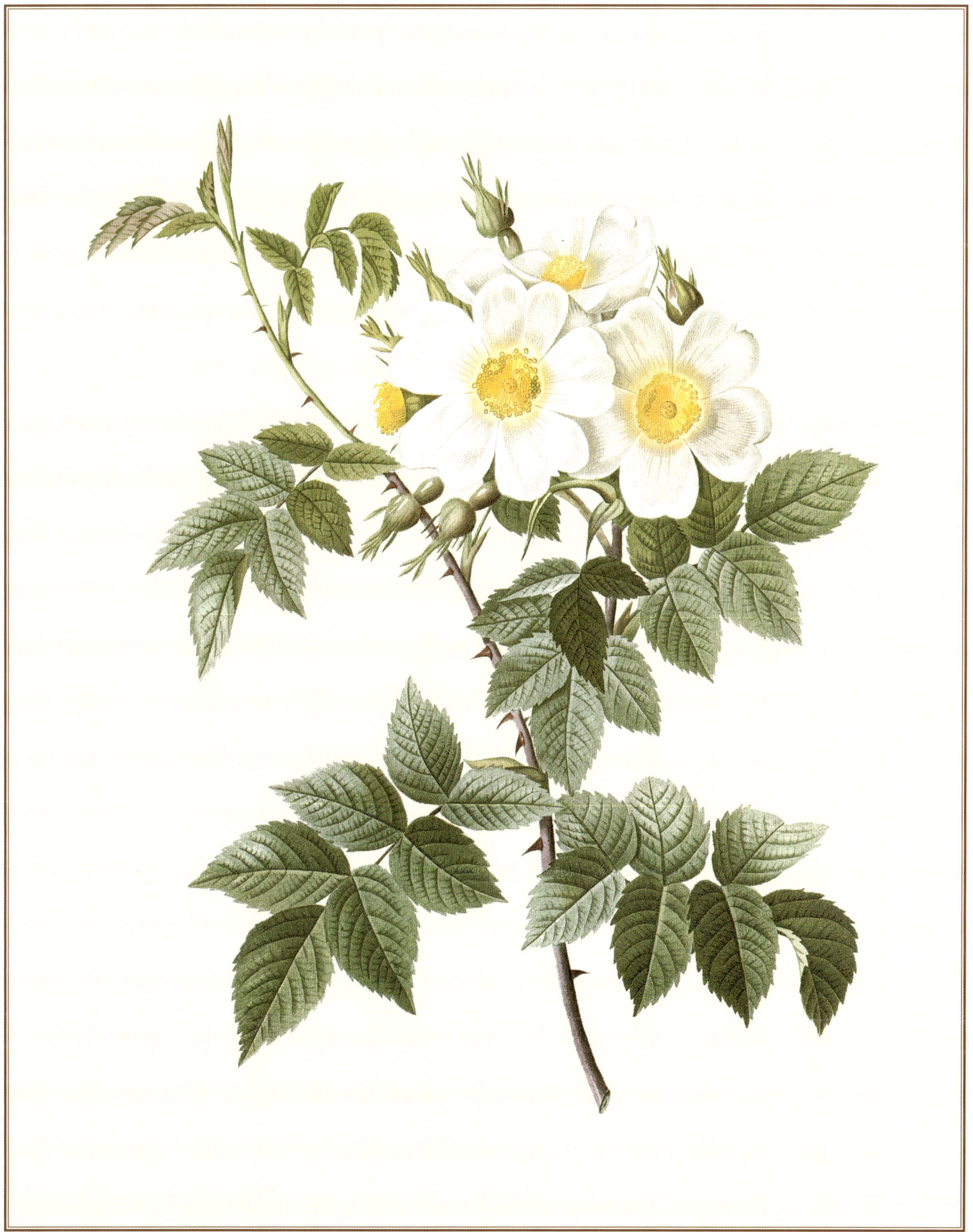

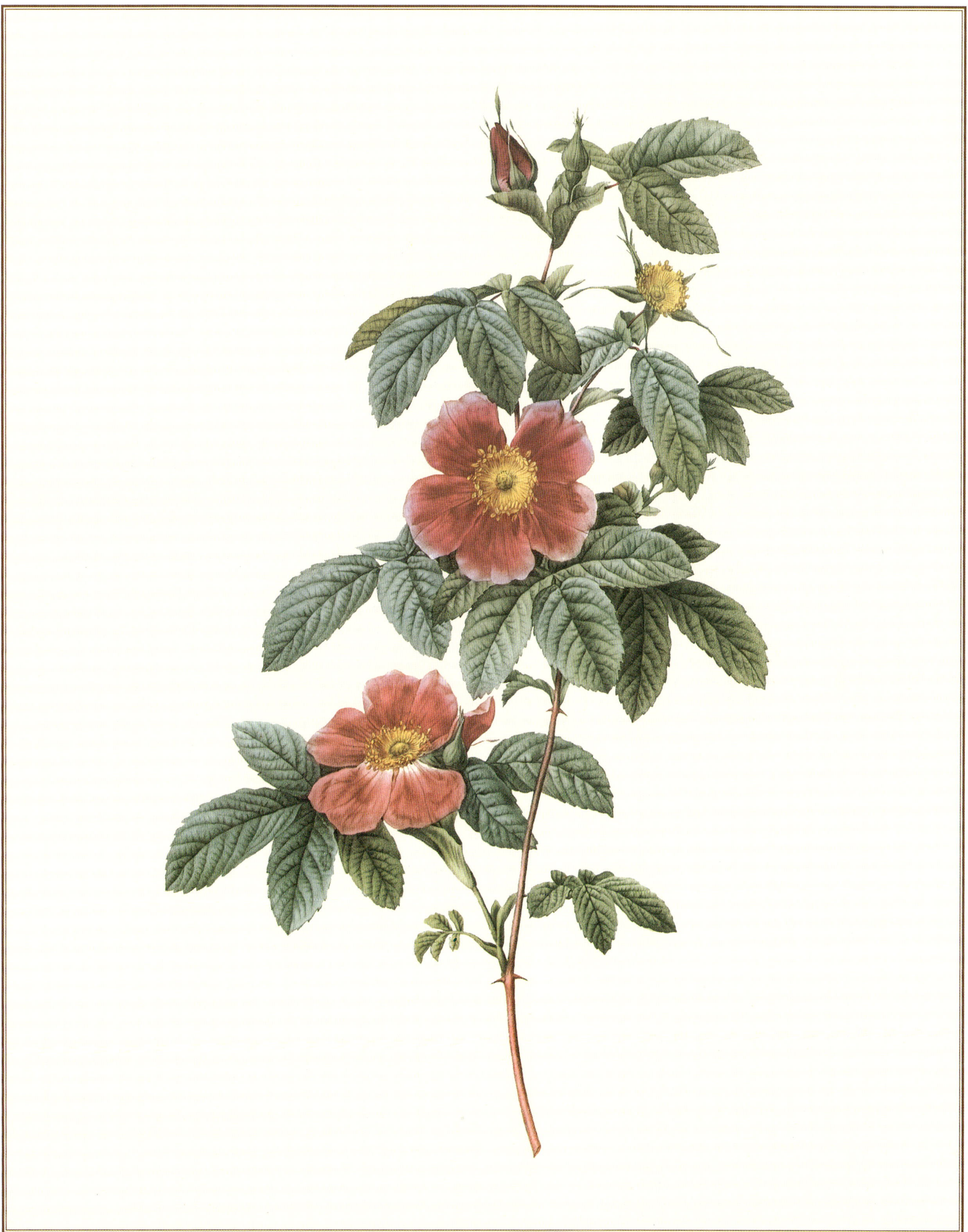

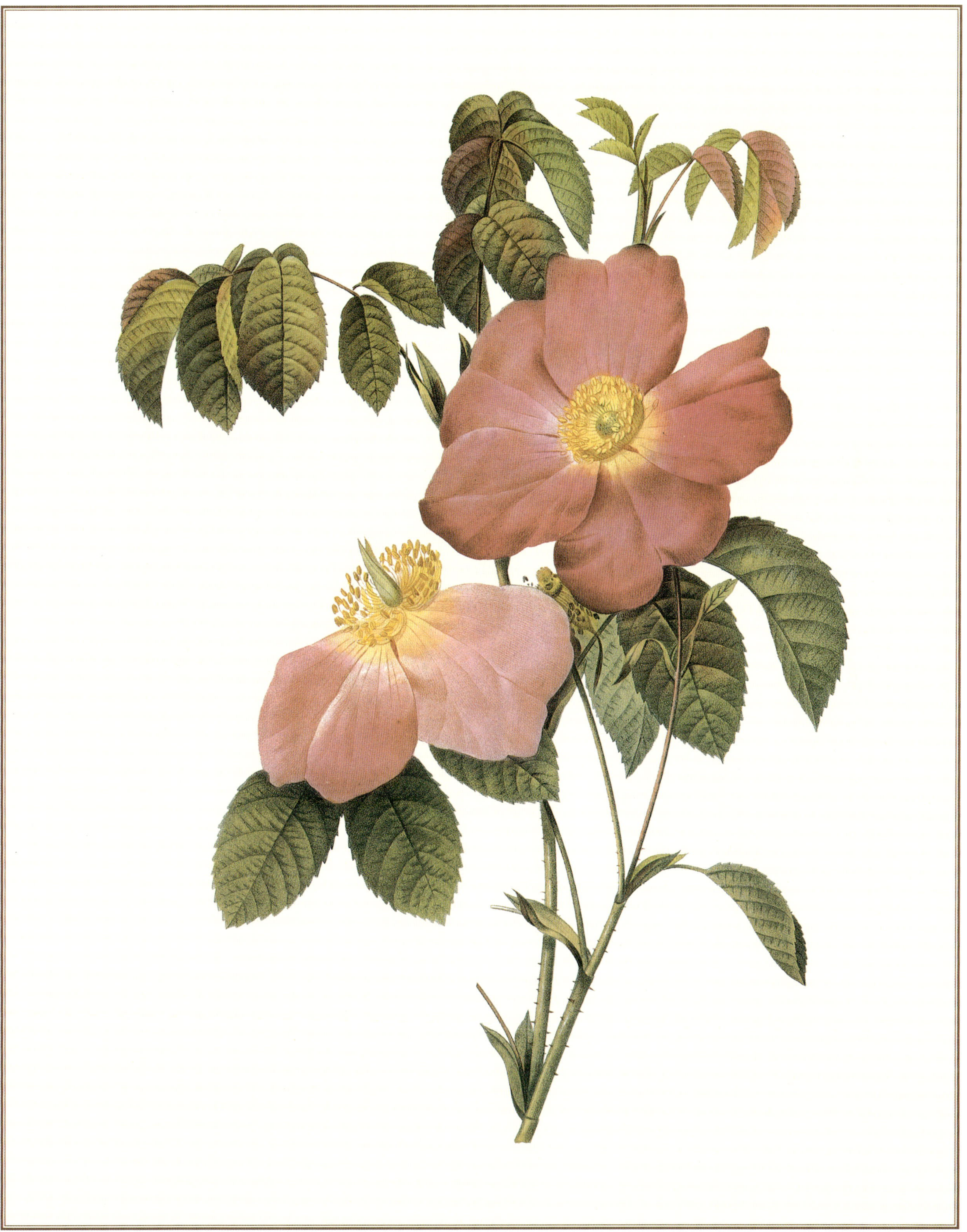

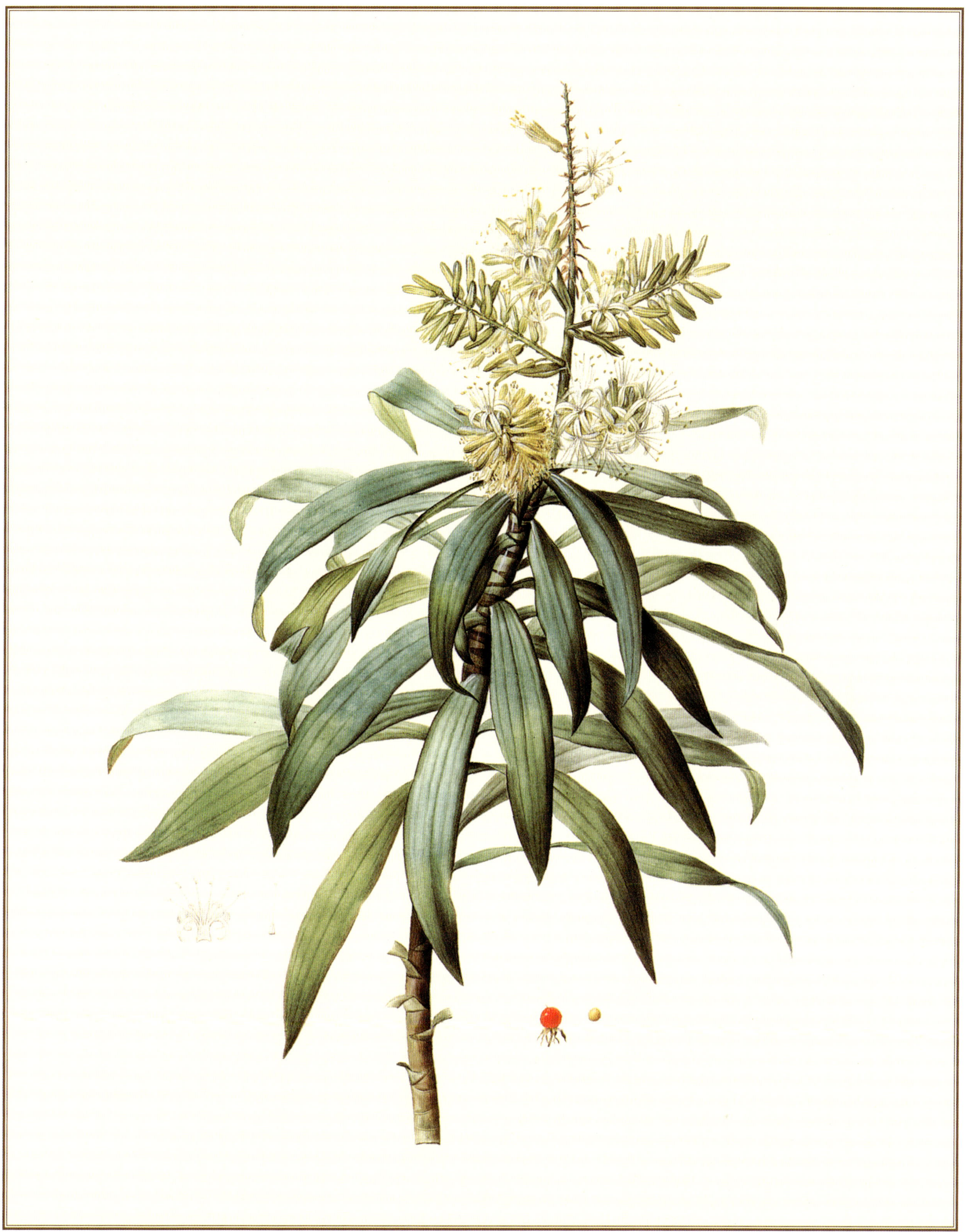

PLATE 74
SPANISH IRIS
(*Iris xiphium*)

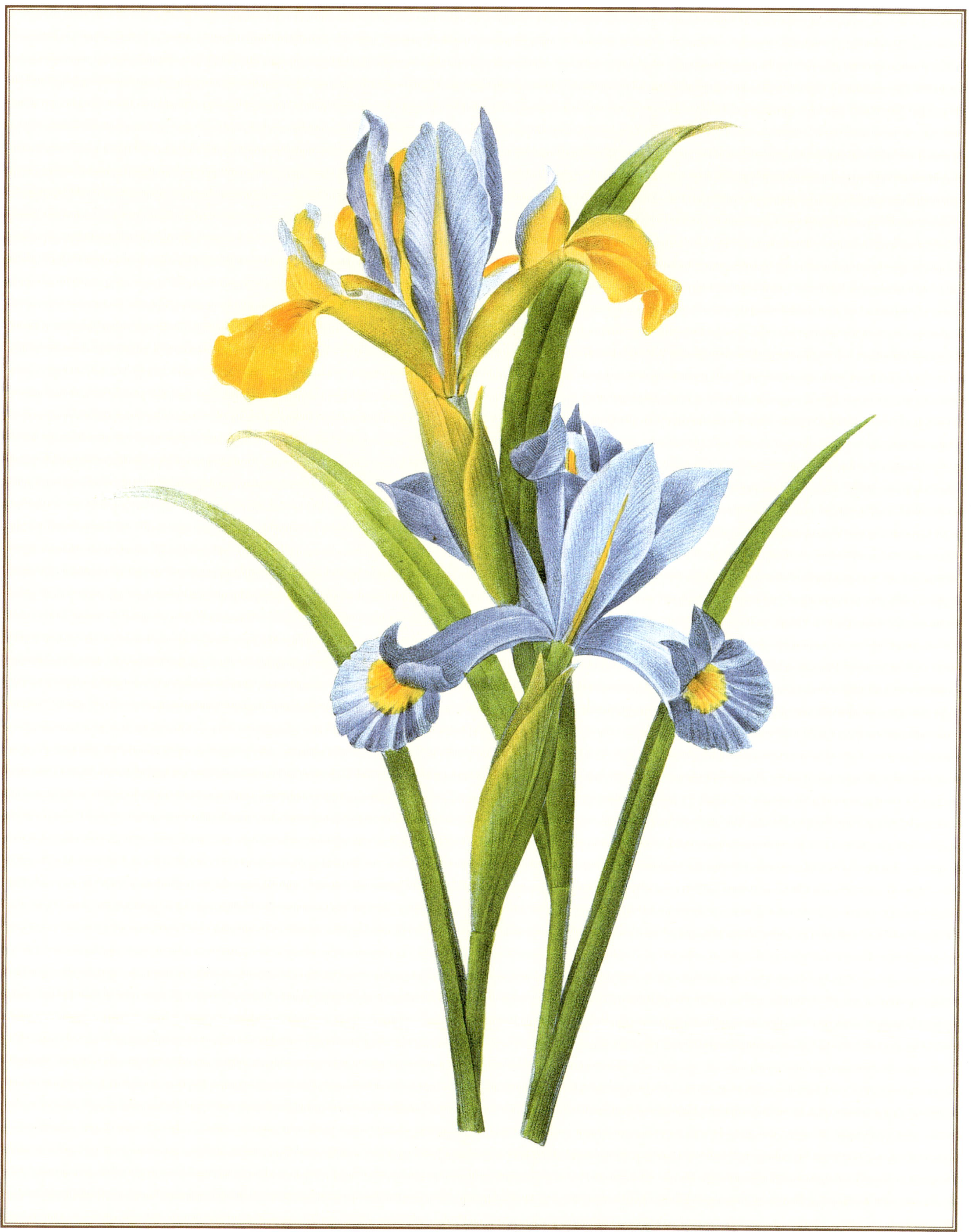

PLATE 75

SPINY-LEAVED ROSE OF DEMETRA

(Rosa spinulifolia dematriana)

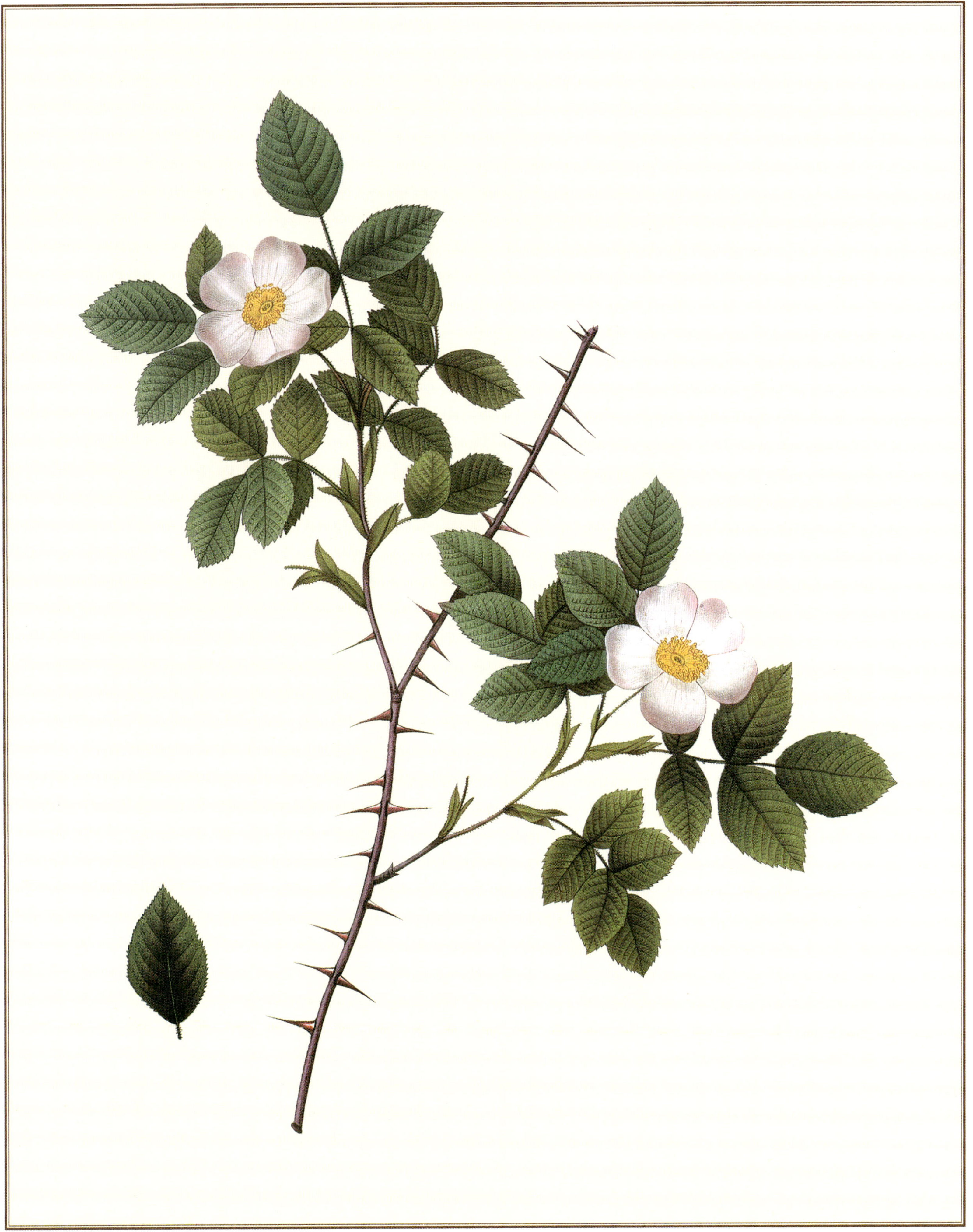

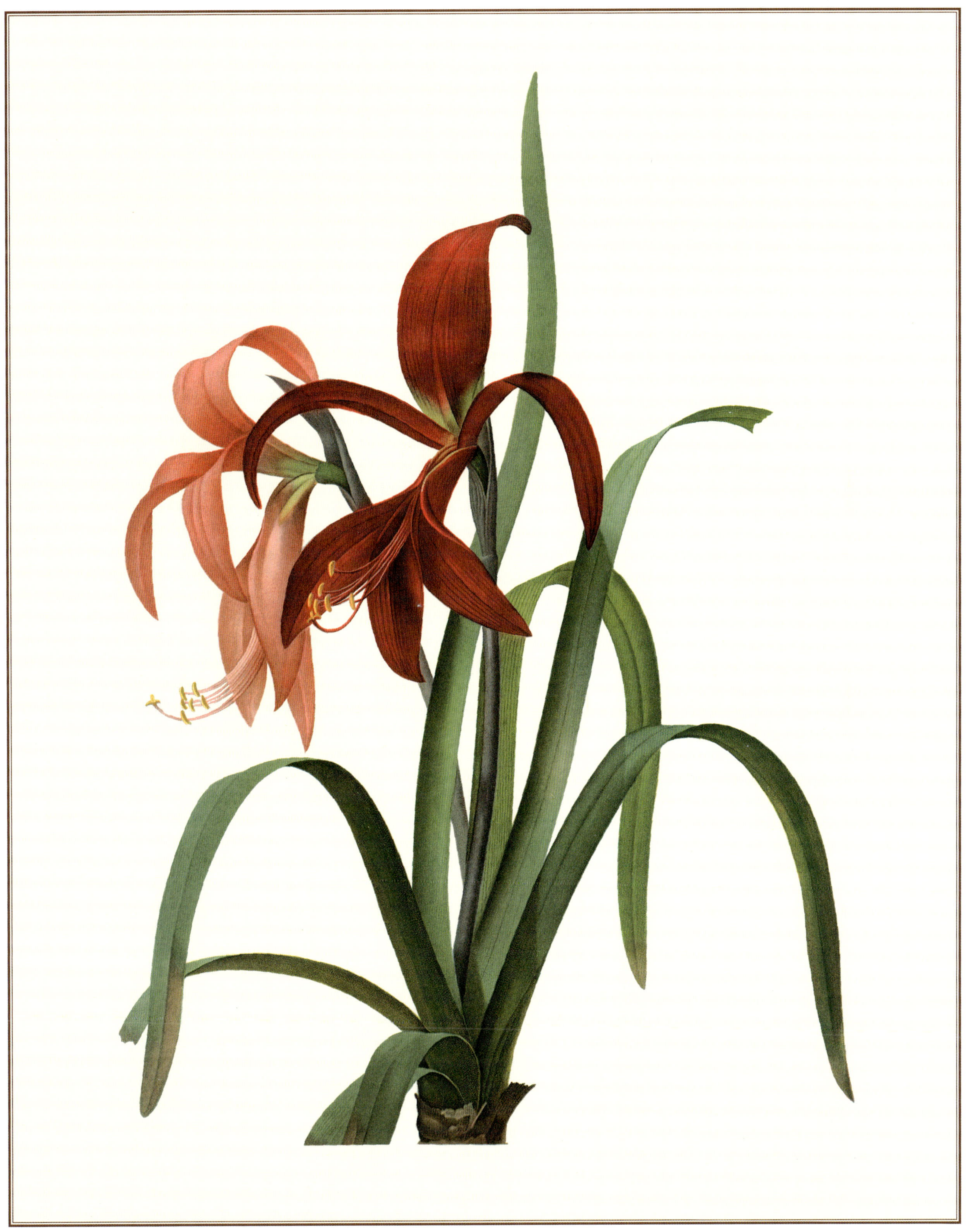

Plate 77
Sweetbriar Rose
(*Rosa rubignosa cretica*)

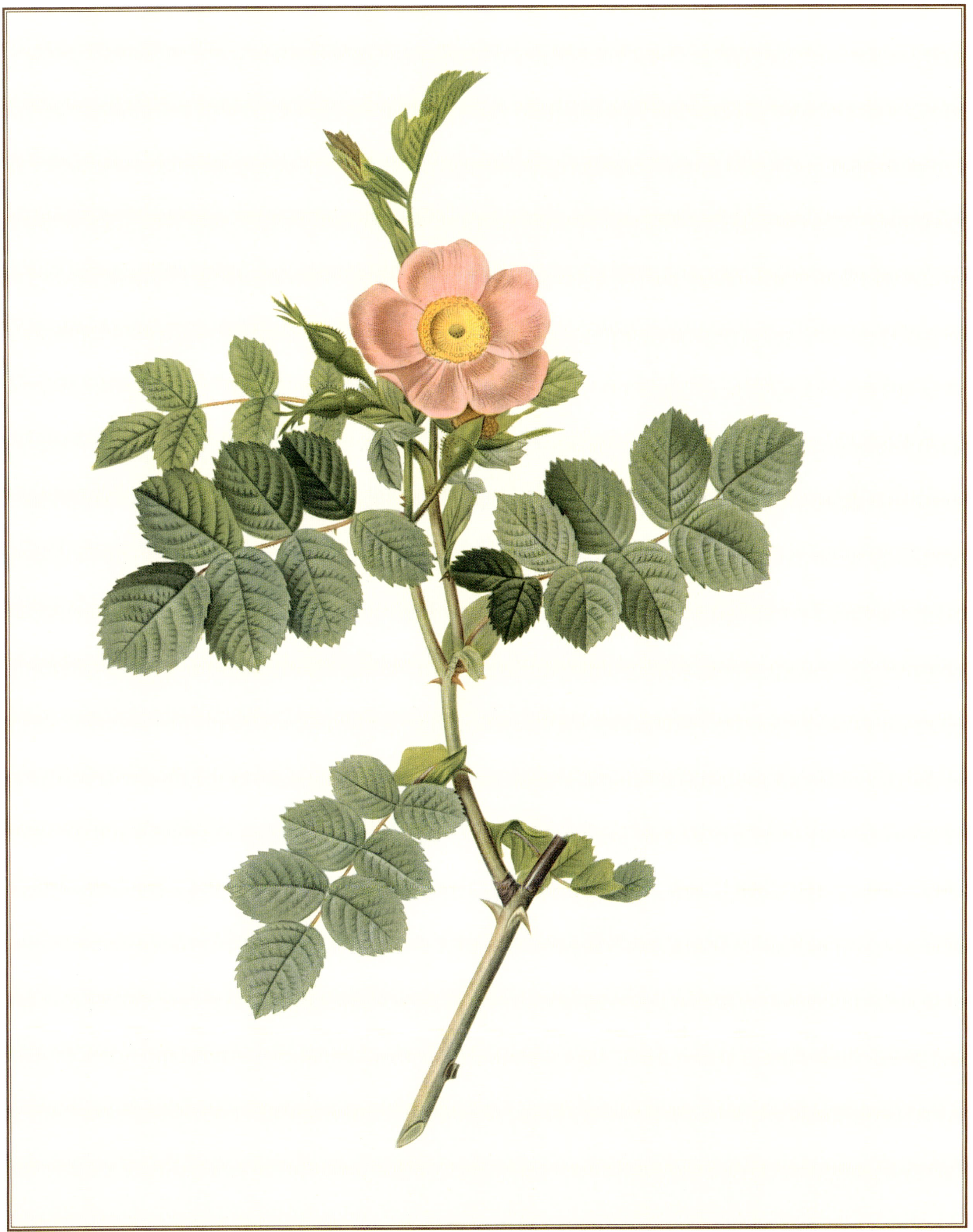

PLATE 78
THORNLESS BURNETT ROSE
(*Rosa pimpinellifolia inermis*)

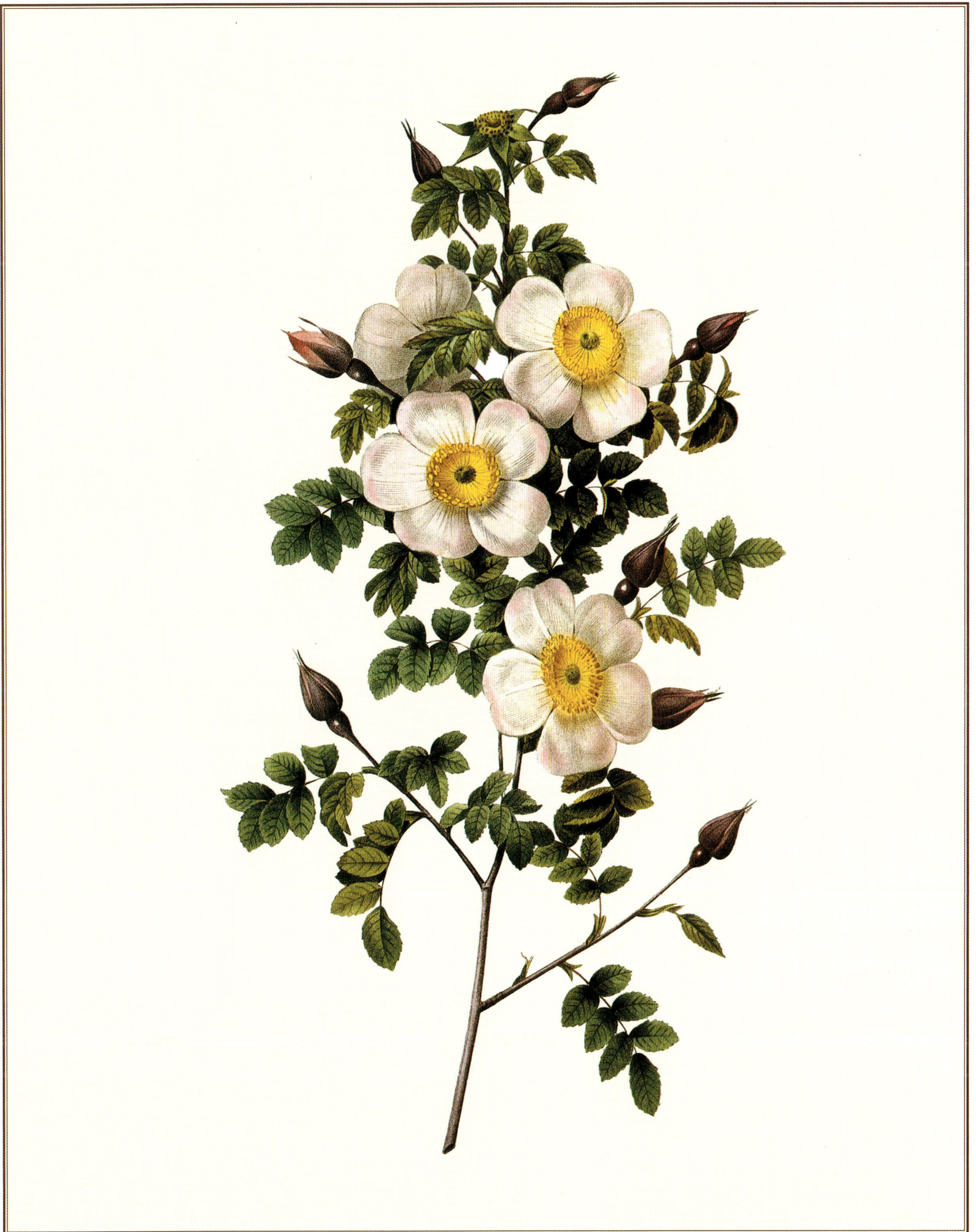

PLATE 79
TIGER LILY
(*Lilium tigrinum*)

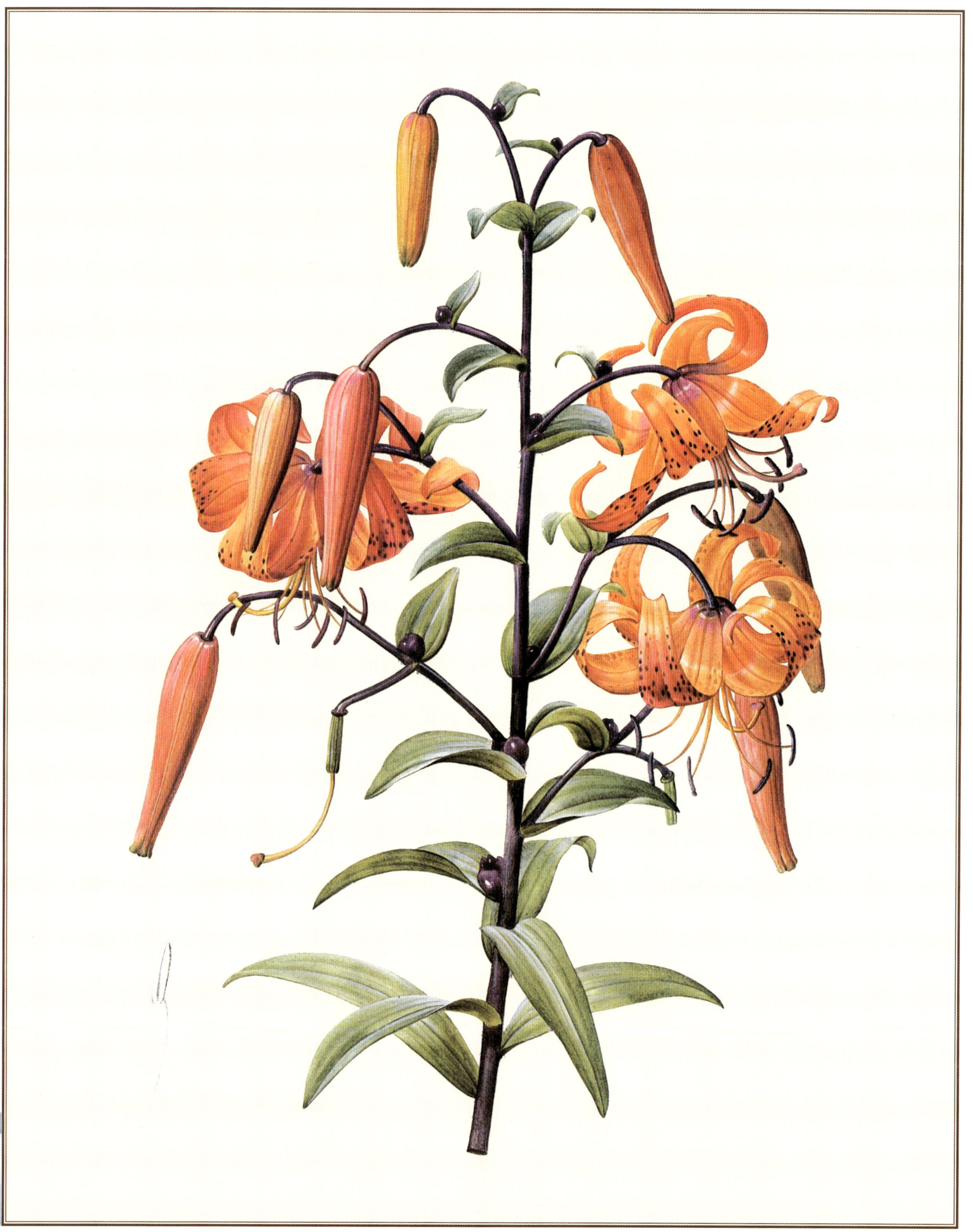

PLATE 80

TUBEROSE

(Polianthes tuberosa)

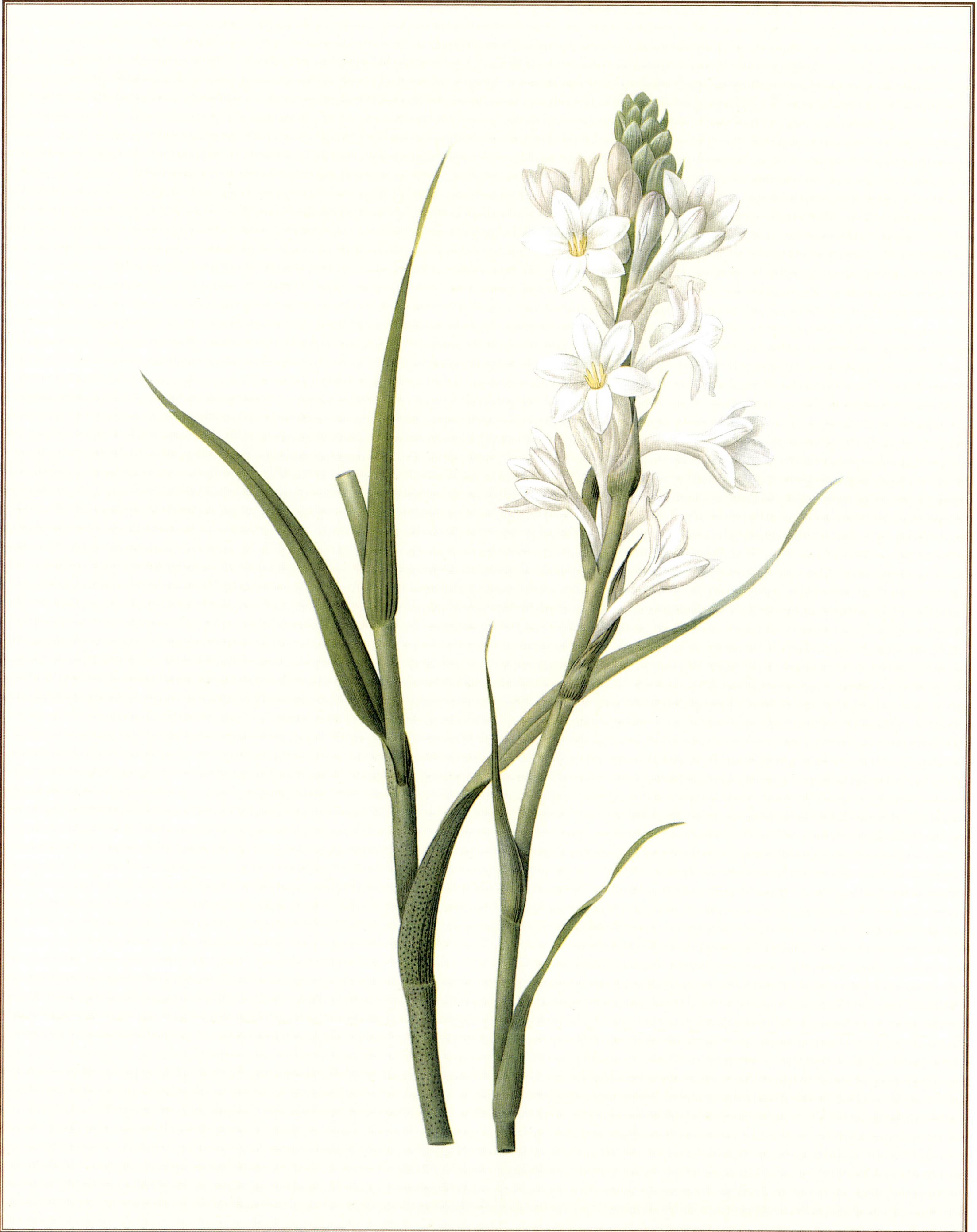

Plate 81

Turk's Cap Lily

(Lilium superbum)

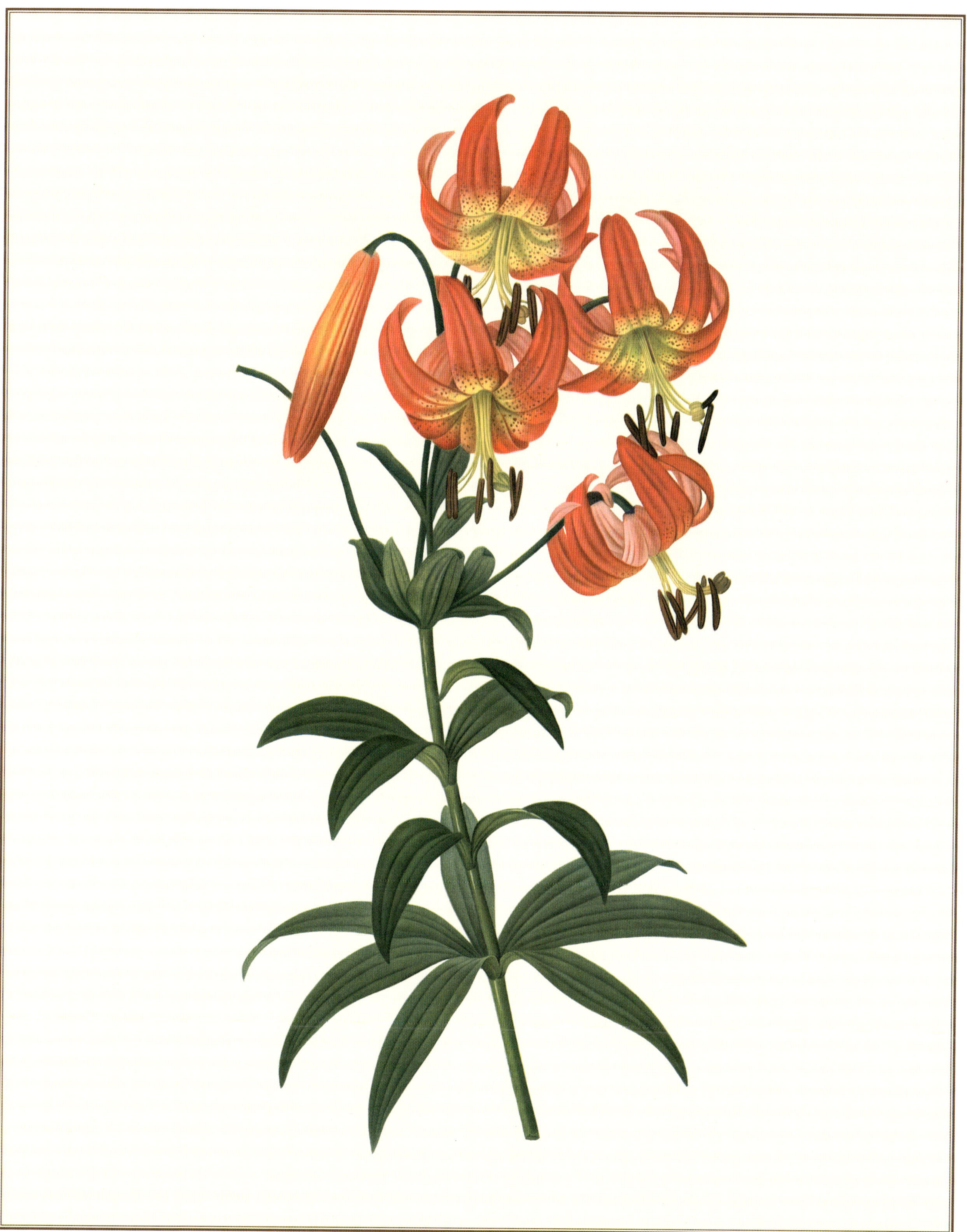

PLATE 82

TWIN-FLOWERED ROSE

(Rosa geminata)

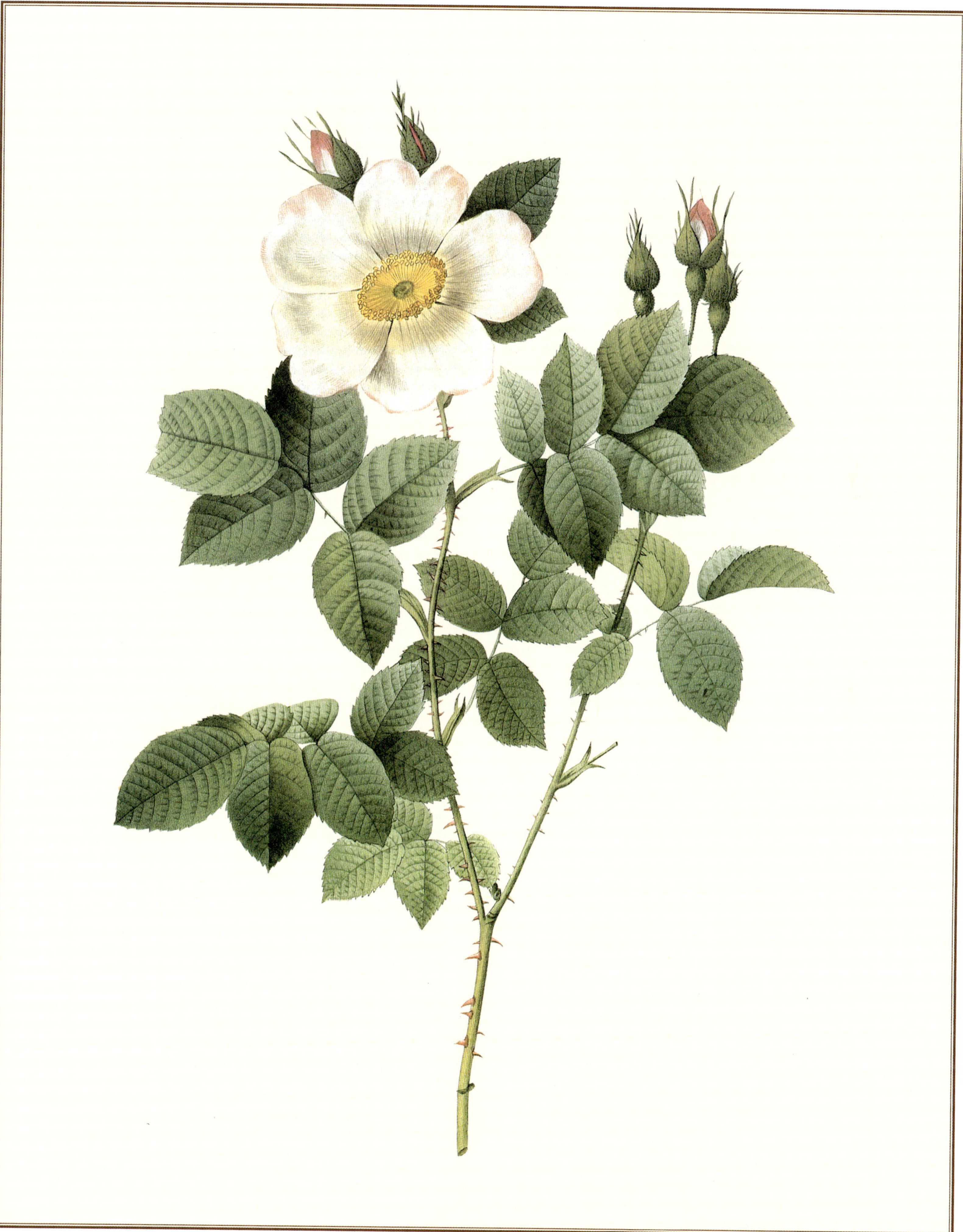

PLATE 83

VAN EEDEN'S ROSE

(Rosa gallica purpurea velutina parva)

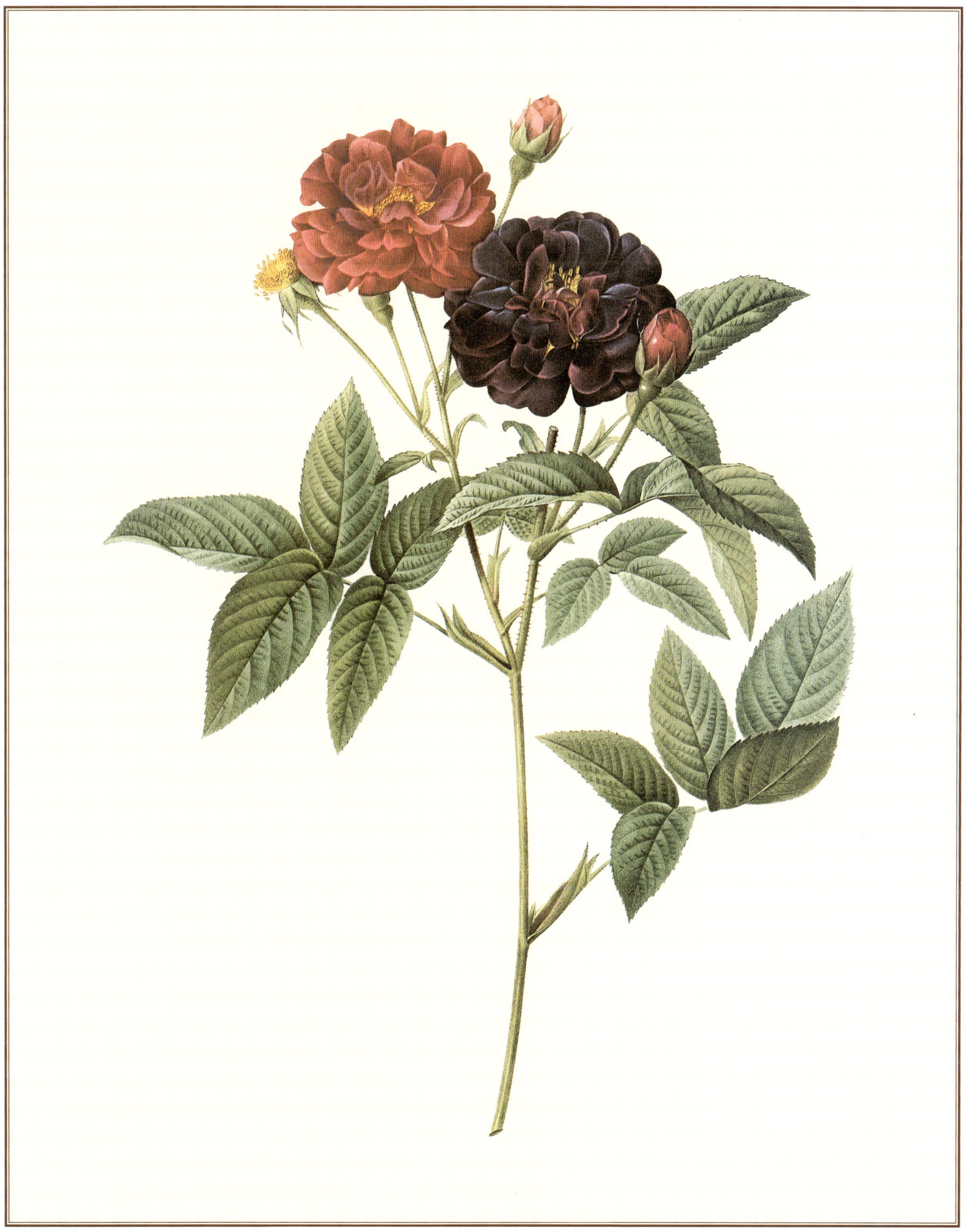

PLATE 84

VARIEGATED ALPINE ROSE

(*Rosa alpina flore variegato*)

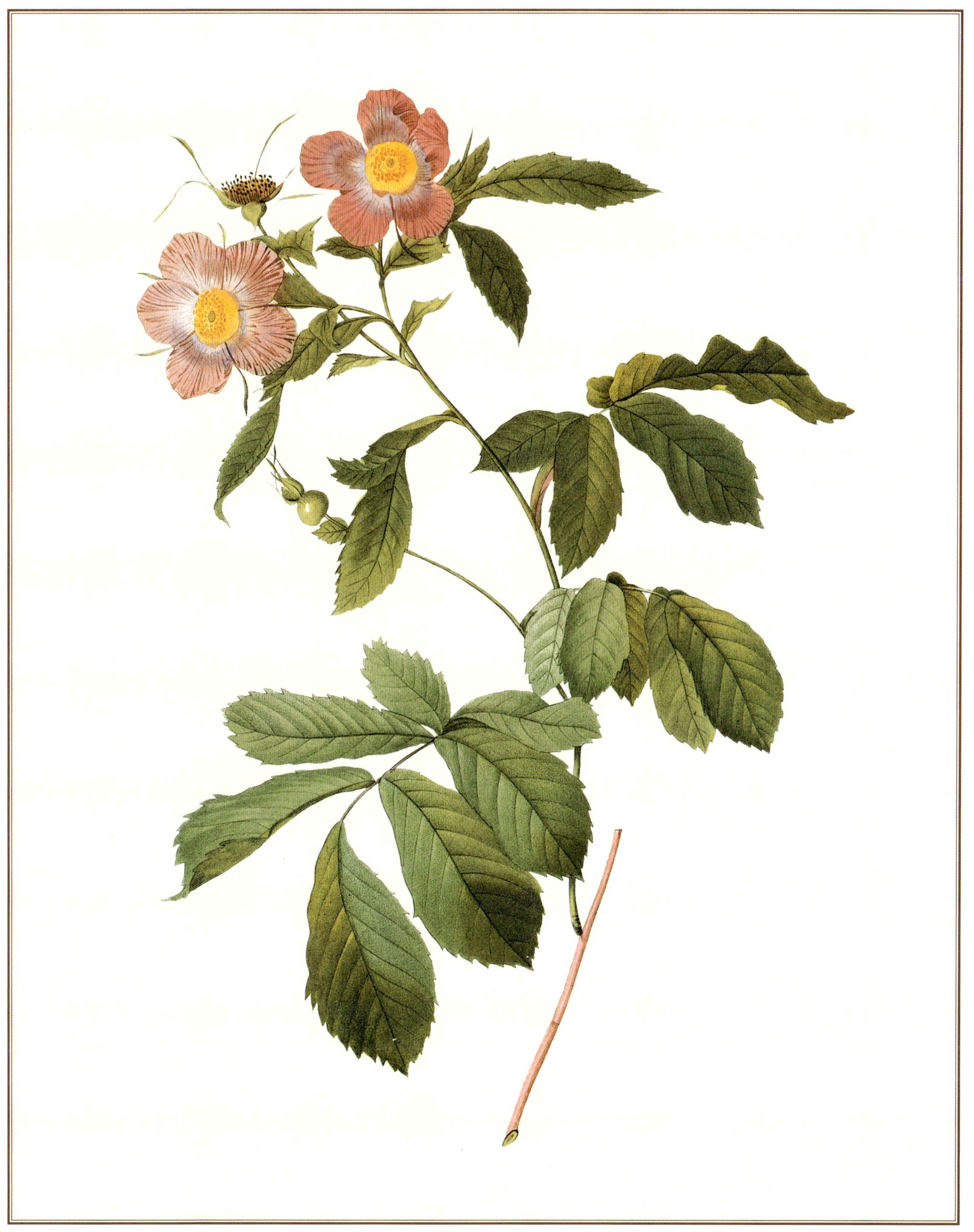

PLATE 85
WALLFLOWER
(Cheiranthus cheiri)

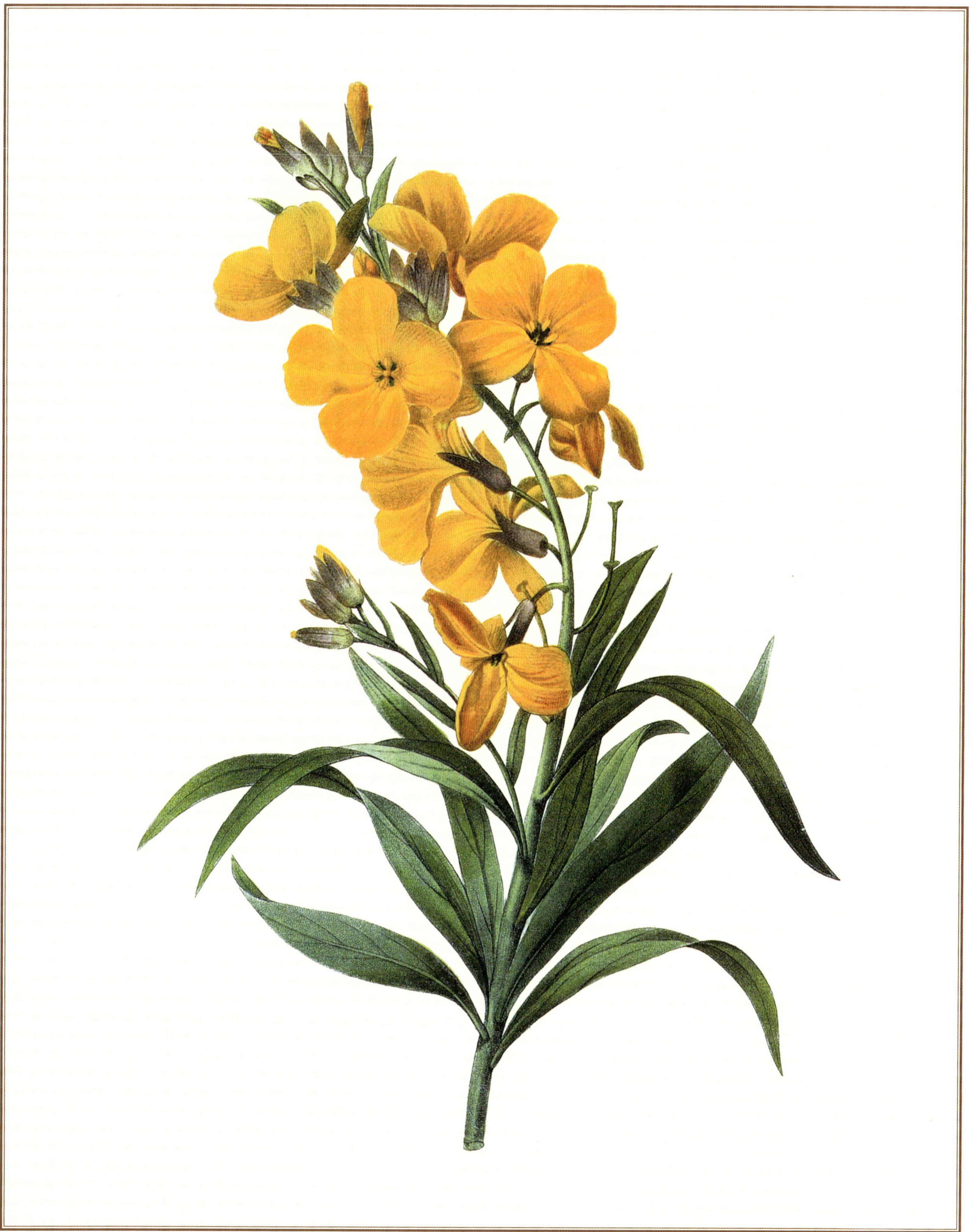

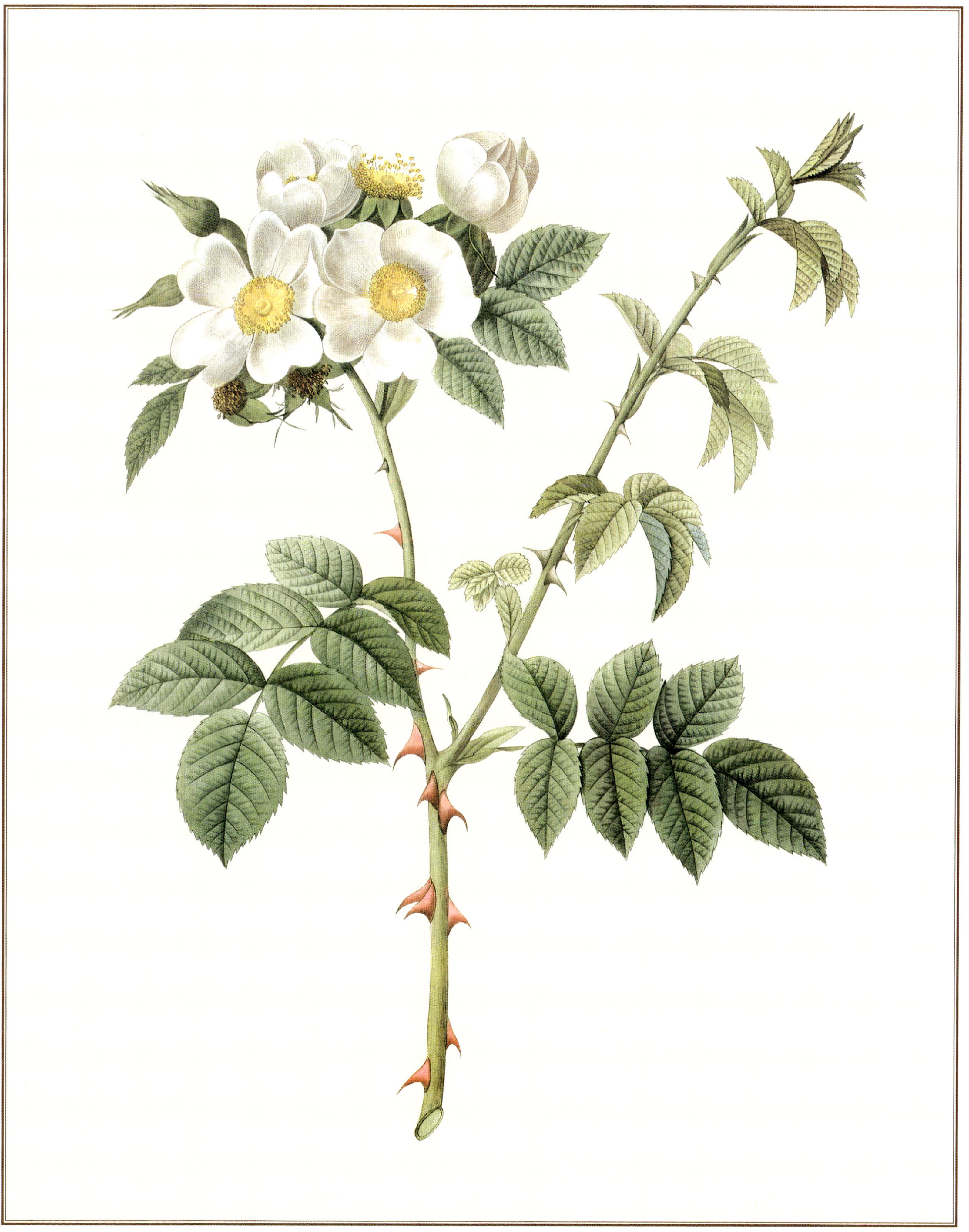

PLATE 87
WHITE MOSS ROSE
(*Rosa muscosa alba*)

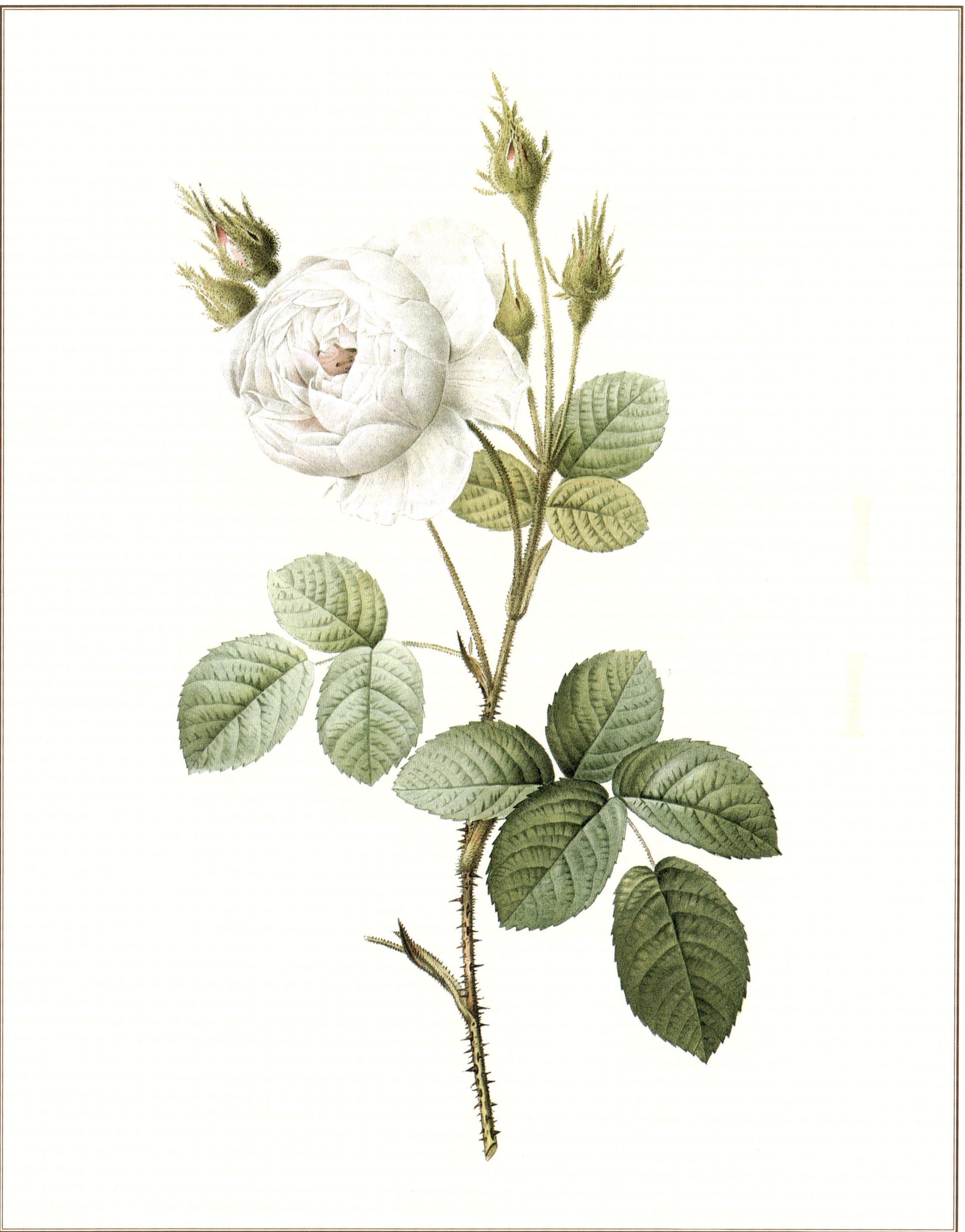

PLATE 88
WINTER'S ROSE CAMELLIA
(*Camellia japonica*)

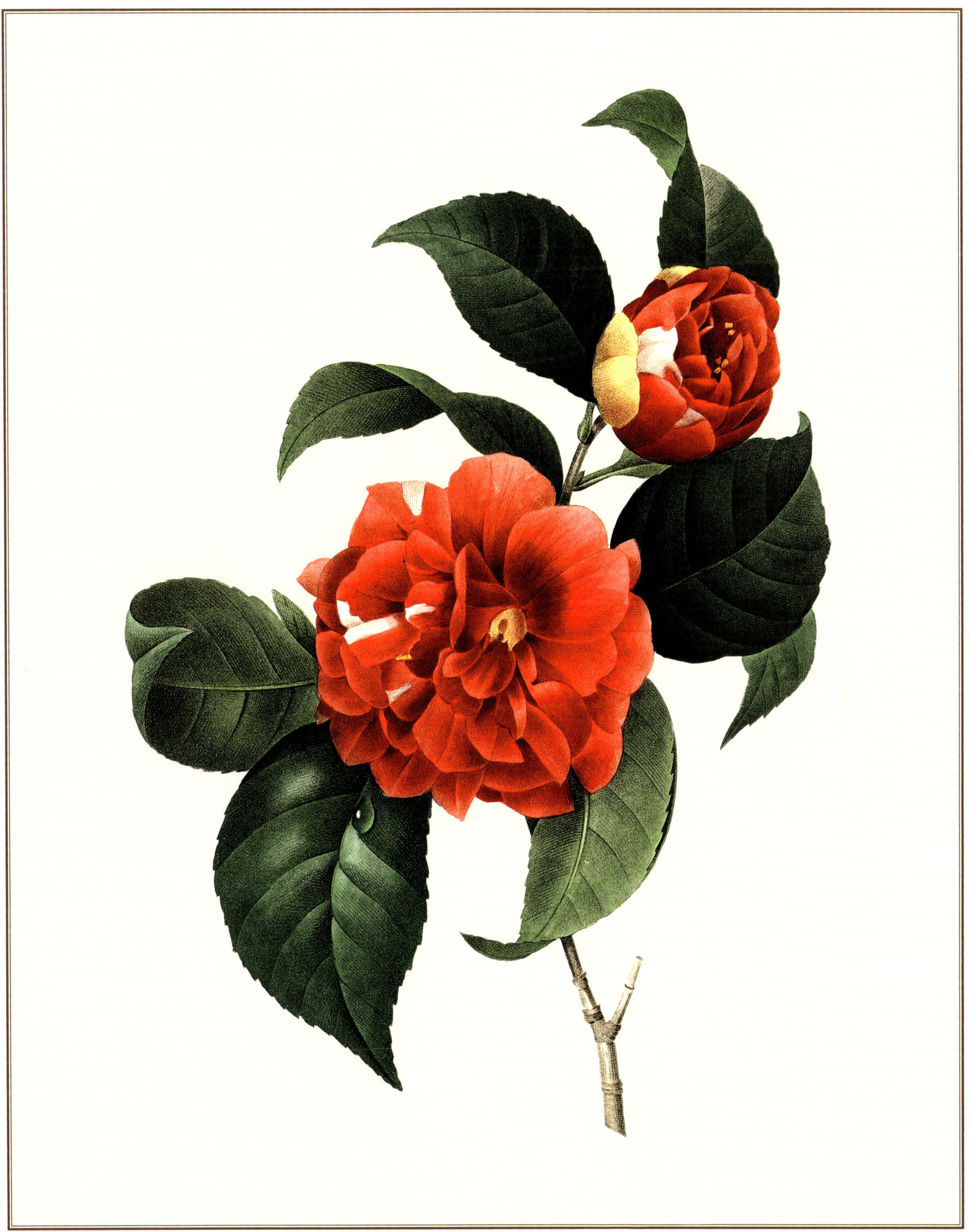

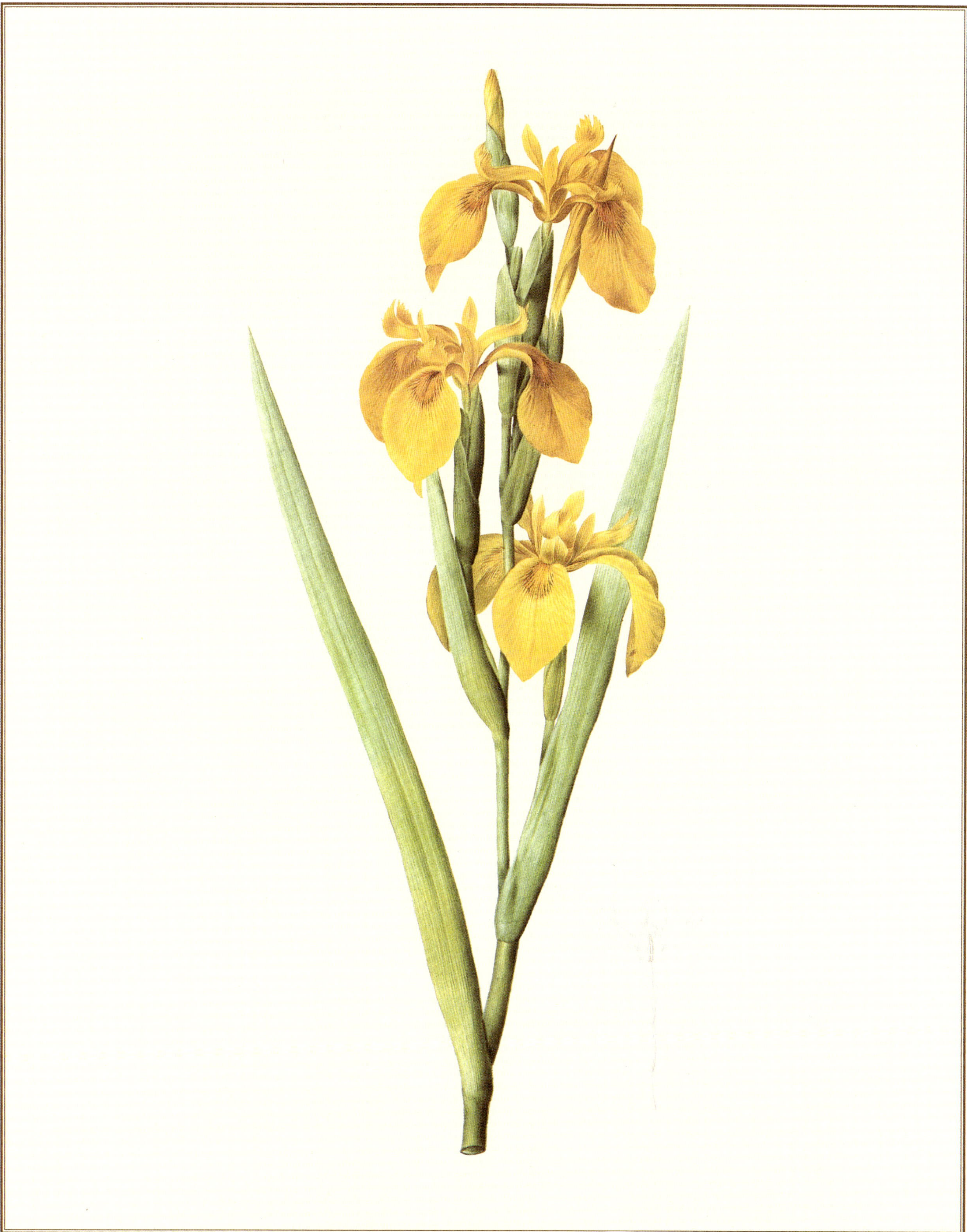

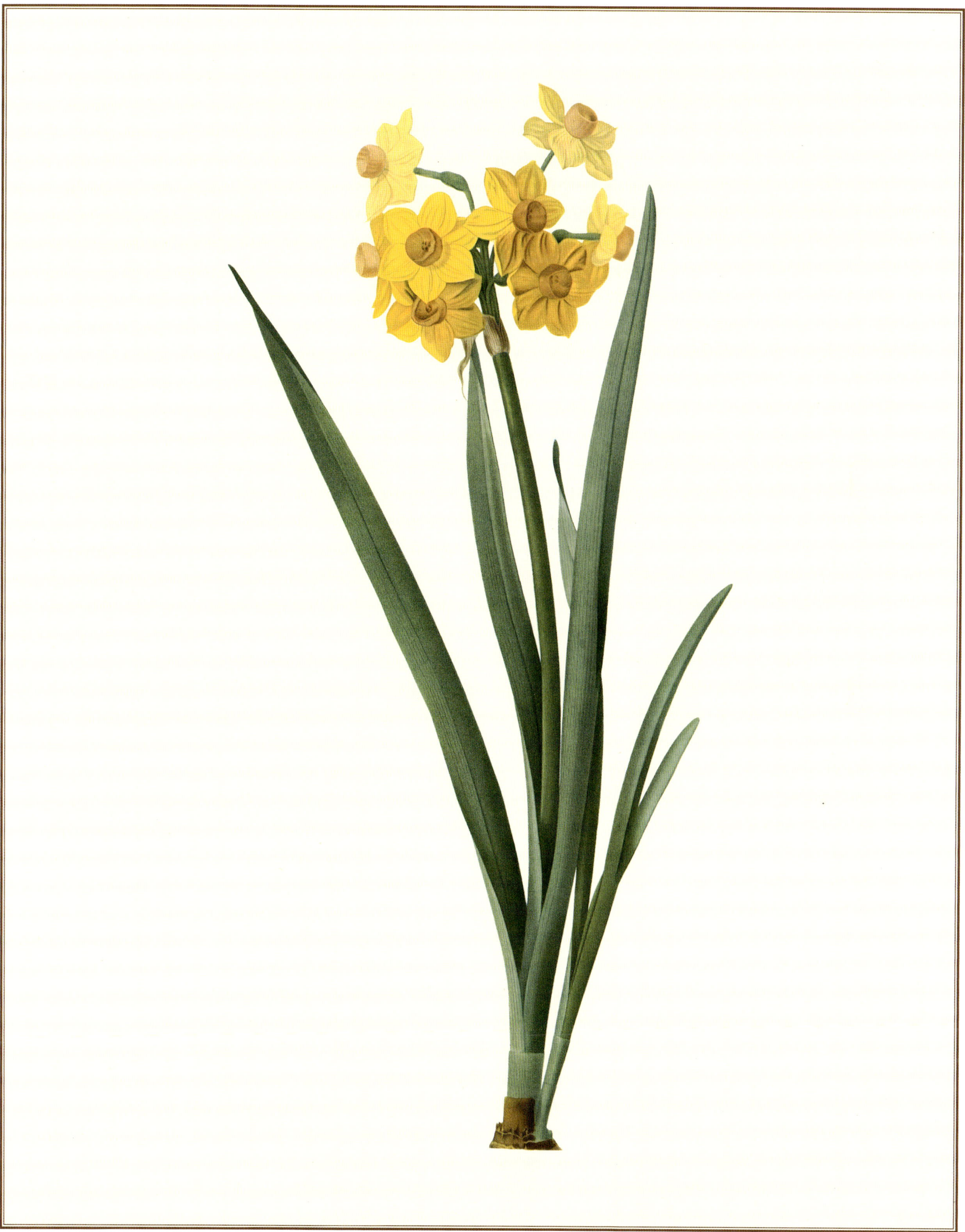